BUILDING IN YOUR BACKYARD

BUILDING IN YOUR BACKYARD

The Suburban Guide to Making Birdhouses, Garden Sheds, Doghouses, Playhouses, Treehouses, Privies, Greenhouses, and Gazebos. **By Victor H. Lane**

Workman Publishing, New York

Copyright © 1979 by Victor H. Lane

All rights reserved. No portion of this book may be reproduced—mechanically, electronically or by any other means, including photocopying—without written permission of the publisher. Published simultaneously in Canada by Saunders of Toronto.

 A Media Projects Incorporated book

Library of Congress Cataloging in Publication Data

Lane, Victor H.
Building in your backyard.

Includes index.
1. Garden structures. 2. Building.
I. Title.
TH4961.L36 690'.8'9 79-12987
ISBN 0-89480-021-3
ISBN 0-89480-020-5 pbk.

Cover and Book Design: Charles Kreloff
Cover Illustration: Jim Trusilo
Book Illustrations: Robert Porter

Workman Publishing
1 West 39 Street
New York, New York 10018

Manufactured in the United States of America
First Printing   June 1979
10 9 8 7 6 5 4 3 2 1

For Caroline,
who is a good
builder.

Contents

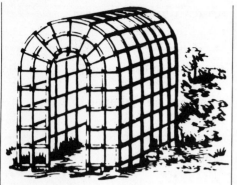

PRIVIES

PLAYHOUSES

TREEHOUSES

GREENHOUSES

GAZEBOS

Introduction

Building anything, be it a loaf of bread or a Taj Mahal, is uniquely satisfying. This is a book about building little houses, from birdhouses to gazebos, that you can put up in your garden or backyard. There are also chapters about doghouses, playhouses, treehouses, garden sheds, privies, and greenhouses. There are lots of pictures, quite a few suggested plans with detailed directions, and a fair amount of commentary and opinion.

I wonder sometimes if perhaps most of the skills and professions practiced by large numbers of people aren't easier than they seem to people who haven't been initiated into their mysteries. After all, anyone of average dexterity can learn to play simple music on the piano, and non-cooks forced to fend for themselves in the kitchen are often surprised to find that relatively little learning and no special talents are required to become a competent cook.

Building is a lot easier than either playing the piano or cooking. Believe me.

All the building projects in this book are within the capabilities of a rank amateur backyard carpenter. The easiest projects are at the beginning of the book, the more challenging ones towards the end. They vary in difficulty from dead easy (see the one-board birdhouses) to moderately involved (the tower gazebo), but all are entirely buildable by someone with very little experience.

In fact, the larger projects don't require higher levels of skill so much as more patience— a longer vision to see the finished product from the time of starting. If you have ever driven a nail or sawn a board, and if you are willing to follow the step-by-step directions I give, you can build any of the little houses in this book.

If you are unsure of your skills, start with a birdhouse. Follow the directions and, above all, take your time. There is an old carpenters' adage (an old saw?) which goes: "Measure once and cut twice; measure twice and cut once." It means, of course, be darn sure of your measurements before you start cutting your wood, but in a more general sense it means, take your time; don't hurry. When you feel a little surer of yourself, have a crack at a garden shed, perhaps, or a playhouse; plans for both are here. When you've done them, you will certainly be ready to tackle a gazebo.

Only a few tools are actually necessary to build most of the little houses in this book: an ordinary carpenter's hammer, a crosscut hand saw, a brace with bits, a plane, a paintbrush, a carpenter's square, a steel measuring tape, and a pencil. Some of the projects do call for a few other hand tools, but nothing exotic.

Power tools are not necessary for any of these projects, but like some of the optional hand tools, they can sometimes greatly simplify a job, or get it done more quickly. A ¼-inch or ⅜-inch electric drill is extremely handy, particularly with a hole-cutting attachment. A portable electric 7¼-inch circular saw (sometimes called a contractor's saw) can do most things a standard hand saw can do, and do them more quickly. Precise cutting is made a lot easier by a radial-arm or table saw, and again, the cuts are made more quickly than you could ever do them by hand. A power jigsaw (sometimes called a saber saw) is great for decorative or fancy cutting, like scroll work. An electric sander is also awfully handy for saving time.

But if you feel moved to build something (and I hope you will), don't rush out and buy a lot of tools in advance, especially power ones. The best time to buy a supplementary tool is at the moment you feel the need for it. If you get really fed up with the effort involved in hand sawing, or if you want to see your project done sooner, that's the time to buy an electric hand saw or, if you have also been having difficulty cutting straight and following your lines, perhaps a more expensive table or radial-arm saw. You'll definitely know when it's time to get a power sander. If you follow this acquisition system you will avoid getting tools you don't need, or ones you find you never use, despite what this or other books may have told you.

If you find building terms like *joist* or *stud* mysterious and vaguely threatening, be reassured. I use them in this book, because the easiest way to talk about joists and studs and the like is to use the names builders have been using for centuries, but at the back of the book is a glossary in which the terms I use are laid out in simple layman's English. Unfortunately, not all the terms are uniformly used in all areas—my *shoe* may be your *sole* or *baseplate*—but at least their usages in this book are explained.

Because many building techniques and processes are used in a variety of different structures, directions for a particular technique or process are given just once, and when you need those directions for another project in the book, you will be referred to the appropriate page. The table of contents also lists all the sets of building directions, so you will be able to find detailed help in building a stud wall, for example, without leafing through the whole book.

I mean this to be a useful book, and an entertaining one as well. I hope you find it to be both.

BIRDHOUSES

The Pleasures of Birds

The prevailing feeling about birds, except perhaps for invasions of starlings and pigeons, is that they are desirable creatures to have around the yard, beautiful in their form and color, in their song, and in their movement. It is a pleasure, I think, simply to watch birds at their various activities, from the time they build the nest to the time they abandon it, the young fledged and off on their own. This is fun for everyone, but especially for kids. Children generally like to watch animals; that's why they like to go to the zoo. But at a zoo, unless you are lucky, it is seldom possible to see animals behaving naturally. At best you can watch them eat, if you've scheduled your visit around feeding time. A family of backyard birds, however, can be observed over a period of some weeks, very much doing their own thing.

Part of the fascination, of course, is simply in watching wild creatures going about their lives under natural conditions. But another part has to do with watching the processes of courtship, "marriage," setting up housekeeping, having young, and sending them out into the world. It is a cycle that birds have, in a broad sense, in common with people, and of which children early become aware that they are a part. By watching a family of birds go through the cycle, particularly if some well-chosen commentary is added by a parent, a child can learn a tremendous amount about what goes on, not just in birds' lives but in people's lives as well.

Then there is simply the pleasure of living in close company with birds. If you're at all lucky, you'll also get some that sing nicely. Listen for the subtle variations in the sounds made by a bird at different times of day, in different seasons; it won't take long before you're able to identify a variety of birds by their songs alone.

If you garden or if you like to sit outdoors in the evening, relatively undisturbed by bugs, an increase in the bird population—if you manage to attract insectivorous birds—can be a very positive boon. Birds eat an enormous amount for their size; small birds may eat as much as their own weight in a day, and that's a lot of insects. Some of the common birds that eat flying insects are swifts, whippoorwills, common nighthawks, hummingbirds, flycatchers, phoebes, martins, swallows, warblers, vireos, scarlet tanagers, cardinals, buntings, finches, and most sparrows.

WHEN BIRDS USE BIRDHOUSES

Birds use birdhouses only for nesting, and they nest only as long as there are young to take care of. Once the young are grown, the nest—whether it be in a birdhouse or in the crotch of a tree—will be abandoned.

Most species of wild birds that nest in birdhouses are solitary, at least during the mating season. Aside from their mate, they will not tolerate other birds of the same species within their territory, the size of which varies according to the species.

IDENTIFYING BIRDS

Unless you are already an active bird watcher, the chances are you won't be able to tell the players without a program. In any good bookstore or public library, there will be a bewildering number of books on birds. To help you select the right ones, several of the standard works are listed below.

Field guides can be frustrating at first. Either the bird you have just seen doesn't seem to be pictured, or, according to the picture and description in front of you, it could be any one of twelve different species. There are several ways out of this muddle. If you have a friend or neighbor who is a bird watcher, get him or her to help. Or find out if there is a local Audubon Society, which may have guided bird walks. Or you can check with your local County Extension Agent. If he isn't strong on local birds, he is most likely to know someone who is. Once you have a good idea what points to look for when you see a bird, you will find the field guides invaluable.

Bull, J., *Audubon Society Guides to North American Birds: Eastern Region,* Knopf, 1977. Although recent, the two Audubon Society Guides have joined the Peterson Field Guides as standard works. Companion to Udvardy, below.

Hickey, Joseph J., *A Guide to Bird Watching*, Dover, 1975. A good book to begin with.

McElroy, Thomas P., Jr., *New Handbook of Attracting Birds,* Knopf, 1960. Helps if you've got an untenanted birdhouse.

Martin, A. C., et al., *American Wildlife and Plants,* Dover, 1951. A lot about feeding habits which can help you attract and keep birds.

Peterson, Roger Tory, *A Field Guide to the Birds* (Eastern Birds), Houghton Mifflin Co., 1968. A standard work. Companion to Peterson, below.

Peterson, Roger Tory, *A Field Guide to Western Birds,* Houghton Mifflin, 1972. Companion to Peterson, above.

Pettingill, Olin Sewall, Jr., *A Guide to Bird Finding (East of the Mississippi),* Oxford Univ. Press, 1951. Useful in telling what birds you can hope to attract in your area. Companion to Pettingill, below.

Pettingill, Olin Sewall, Jr., *A Guide to Bird Finding (West of the Mississippi),* Oxford Univ. Press, 1953. Companion to Pettingill, above.

Udvardy, M. D., *Audubon Society Guides to North American Birds: Western Region,* Knopf, 1977. Companion to Bull, above.

Some species, however—notably starlings, purple martins, and doves and their relatives—like to nest in groups. It would be wrong to say of these gregarious birds that they are not territorial; they are, but their territories are much smaller than those, say, of robins, who commonly demand as much as half an acre to themselves. Martins, for example, live together in quite large numbers. Close-nesting flocks numbering in the tens of thousands have been observed in the wild. But in a martin house, each pair must have its own compartment, with private entrance. If another martin tries to horn in on an occupied apartment, all hell breaks loose.

The fact that birds are territorial does not mean that you have to settle for doves or martins if you want to have more than one pair of birds nesting near your house. Most birds protect their territories only against birds of their own species; birds of other species can often nest close by without any problem.

From the time birds abandon their nests until it is time either to migrate—or, for birds that don't migrate, time to start nesting again in the spring—the birds will roost at night wherever

The variety of materials from which you can make birdhouses is almost infinite. The one on the left is made of tarpaper, glued together. The one on the right is of birchbark.

they feel like roosting. Roosting patterns, even for quite common birds, are not well understood—in part, at least, because roosting is done at night when even seeing birds is difficult. Some birds appear to favor the same roosting site night after night; others seem to prefer frequent changes of location. So you may or may not see members of the family you watched during the nesting season once that season has passed.

Some birds, as it happens, won't nest in a birdhouse at all,

so you aren't likely to attract them as residents. But the U.S. Fish and Wildlife Service has identified a number of species that are known to have nested in birdhouses or on nesting shelves, and a list of them can be found in the chart on page 27. Given a choice, most birds will select a birdhouse of the proper size over an open nesting shelf; they value privacy.

But some will use a nesting shelf if nothing better presents itself, particularly if the shelf is in a secluded, quiet place. That is

This product of a 19th century Russian woodcarver is a birdhouse for starlings—a bird most Americans would not go to this kind of trouble for. The birds enter through the mouth and nest in the head. The sign the woman is holding means, "to sniff." I have no idea why.

nice for the birdwatcher, as birds on a shelf are easier to watch than birds in a house.

ATTRACTING THE BIRDS YOU WANT

If you like to see wild birds around your house, there are some useful things you can do to increase their number. According to Roger Barton in his book *How to Watch Birds*, (New York: McGraw-Hill, 1955), the government has estimated that there is an average of one bird per acre in the eastern United States, but that in suburban areas there can be as many as ten birds per acre if people see to it that water, food, and nesting sites are made available for them.

Attracting birds is a little like fishing—except that the birds generally come off better than the fish do. You can either set out food, put up birdhouses, and take what comes, or you can tailor your efforts in such a way as to attract predominantly one or two species of birds. If you adopt the former course, the chances of developing a bird population in your yard are good, although it is also likely that birds will be the most common ones, such as house sparrows or starlings.

If you decide to be selective in what you try to attract, then you have to know a little about what you're doing. One thing you must find out is whether the species you are interested in frequents the area where you live. It's sad, but there is absolutely no point in a resident of Katonah, New York, setting his heart on attracting a golden cockatoo or a hairy emu. It is also not enough for our man in Katonah to know that a bird he wants to attract is known to frequent southeastern New York State, where he happens to live. Some birds simply will not nest close to a house, others require an immediate environment containing specific kinds of vegetation or a body of water.

A good way to attract any species is by putting out the food that it likes. (A food chart for various species appears on page 20.) Keep the birdbath filled and make sure that both bath and feeder are well protected from predators—particularly cats. (For predator guards see page 40.) If the feeder and bath are well stocked before the nesting season begins, they will help attract birds to your birdhouse. If you want to have bluebirds nesting in your yard, but none ever come to the feeding station, it would be foolish to count too much on attracting a pair by putting up a bluebird- tailored birdhouse.

Unless you are more than usually Franciscan in your outlook, you don't put out and stock a feeder just for the birds. You do it for yourself, too, because you like watching them while they eat. Obviously, then, you must put the feeder somewhere where it can easily be seen from the house. An ideal place is by a window near where the family has their meals. But—and this is a big but—if you have small children who would find it difficult to stay quiet when a bird was joining you for dinner, you'd do better to mount the feeder on a post 15 or 20 feet from the window.

Particularly nice feeders are the ones often sold in hardware stores, with glass walls on three sides, a roof, and perches extending out from the open front. They keep the worst of the weather off the feed, allow you to see the birds, and give those birds, who like that sort of thing, a twiglike landing place.

Or you can go the other route and make your own feeder. A

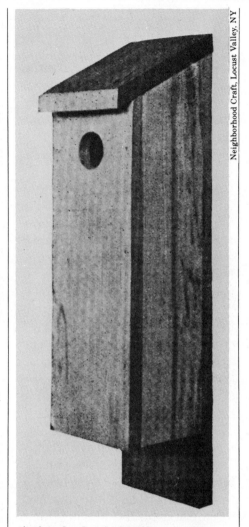

A simple, basic birdhouse. The top hinges open for cleaning.

Neighborhood Craft, Locust Valley, NY

good universal feeder can easily be made with a platform about 12 by 18 inches, walled on two short and one long side, with a slanting roof, 8 inches high over the open long side, and 4 inches high at the closed long side. If the walls on the short sides are made of glass or plastic, it will be easier to watch the birds as they eat. One or two landing dowels, about 9 inches long and about ½ inch in diameter, should protrude from the front of the platform, and on the open side there should be a low lip attached to the edge to keep the food from slipping off. The walls and roof are not necessary, but they help keep wind and rain from driving seed and other food off the platform and onto the ground.

The food will seem safe to birds if the feeder is protected from predators. In the main this means cats, but there are other enemies as well. As it happens, birds aren't the only creatures that like birdseed. Squirrels love it too. Some birds—bluejays, for example—are well up to competing with the squirrels for food, but most other species aren't. So put up a functional predator guard on the feeder.

Pet stores and supermarkets usually sell a birdseed mix that has something for everyone,

FOODS FOR SOME BIRDS

Birds can be attracted by a lot more than seeds and suet. Of course, most foods are enjoyed by a number of species, so keep an eye on the feeding station and see what arrives.

Here is a list of some of the foods particular birds like. And the list is by no means exhaustive. There is hardly a kind of food scrap that some bird won't like, but remember to clean perishable food off the feeder before it begins to decay.

Blackbird	Sunflower seeds, corn, cracked nuts, scratch feed (commercial chicken feed)
Bluejay	Almost anything, but favorites are sunflower seeds, suet, cracked nuts, corn
Bobwhite	Suet
Cardinal	Sunflower seeds, melon seeds, corn, cracked nuts, scratch feed
Catbird	Apples, oranges, currants, raisins, breadcrumbs
Chickadee	Small seeds, sunflower seeds, peanut butter, cracked nuts, breadcrumbs, suet
Finches	Sunflower seed, millet, scratch feed, wheat, small-seed mixtures, breadcrumbs
Goldfinch	Canary seed, millet
Hummingbird	Sugar water—one part sugar, two parts water. (A hamster feeding bottle, available in pet stores, makes an excellent hummingbird feeder. Hang it just outside a window. Change the water every couple of days to keep it from fermenting in the sun.)
Junco	Mixed small seeds, scratch feed, wheat, breadcrumbs, millet
Mockingbird	Cut apples, currants, raisins, breadcrumbs
Mourning dove	Mixed small seeds

Nuthatch	Suet, sunflower seeds, peanut butter, cracked nuts, breadcrumbs
Pheasant	Cracked corn
Purple finch	Sunflower seeds, hemp seeds, millet, chick feed
Redpoll	Currants, raisins, breadcrumbs, cut apples and oranges, short lengths of cooked spaghetti
Sparrow	Mixed small seeds, scratch feed, millet, wheat, breadcrumbs
Starling	Suet, cut apples, currants, scratch feed, table scraps
Thrush	Cut apples and oranges, currants, raisins, breadcrumbs
Titmouse	Suet, peanut butter, cracked nuts, sunflower seeds, breadcrumbs
Towhee	Scratch feed, cracked nuts, corn
Woodpecker	Suet, cracked nuts, corn

Although a bird feeder might attract birds to your yard, it is not a bad idea to discontinue feeding once nesting has commenced. Birds should forage naturally whenever possible, and the food they find that way may well be healthier for the nestlings in the long run than what you might provide on the feeder. Furthermore, birds fed by you will be less active in controlling bugs and insect pests in your area.

Some bird experts also discourage feeding birds during the summer; they feel it makes the birds dependent on people's largesse and less able to forage for themselves if that largesse should cease, or if, as is likely, they fail to find the free lunch in the place to which they migrate, come fall. Other experts believe that this is not a problem, that the only mortal effect summer feeding is likely to have is to prolong the life of an old bird that might have difficulty foraging, but yet be able to survive a bit further into old age with the aid of a feeding station. I know of no hard experimental evidence to support either point of view.

which is fine. If, however, you want to feed selectively, then have a look at the food chart. Remember, birdseed, no matter what the mix, is only going to attract seed-eating birds. Lots of birds eat insects and grubs and the like; others are partial to fruit. The insectivorous birds will appreciate a suet ball. Many supermarkets sell suet balls, but if yours doesn't, you have only to wrap some beef fat up with cotton thread and hang it up. A ball about the size of a softball is just right. People commonly hang suet balls from a tree limb, although if you do that, it's nice to give the birds something to hang onto while they peck. If you hang the ball in one of the small plastic mesh bags that super-markets package produce in, the birds should be perfectly happy. But don't use a red bag or red string to hang up the suet. Many birds don't like red.

Birds need water just as much as they do food. All year round they need it for drinking and bathing; in the nesting season some birds need it to make mud for plastering their nests. The most critical times, as far as your intervention into the birds' lives is concerned, are in winter, when familiar sources have frozen, and during drought periods any time of the year. The best way to provide water is probably in an ordinary birdbath. The ones made of concrete are particularly good because the rough surface is easy for birds to stand on, and the dish part can be lifted off for emptying and cleaning without a great deal of fuss. An unexpected bonus of this kind of bath is its surprising cat-proofness.

A bath requires frequent refilling and occasional cleaning, but simpler alternatives sometimes present themselves quite naturally. One very dry summer when we lived in Nebraska I simply allowed the garden hose to drip slowly—a drip every couple of seconds was ample to create a mud puddle that never got very large, but never dried up either. It was a big hit with the birds.

LOCATING A BIRDHOUSE

Many of the same caveats apply to locating a birdhouse as locating a feeder. Birds want privacy when they nest, and will avoid any site that seems too public, either to people or other animals.

Of course, if you are putting

Although this martin-house colony is made from hollowed-out gourds hung from rough boards on an even rougher pole, the sum effect is lovely and not crude at all.

up a birdhouse simply to be nice to the birds, the more secluded you have it, the more likely it is to appeal to birds. Birds will be uneasy and perhaps not nest in a birdhouse placed close to a busy road or near a frequently used walkway or, say, a children's play area. On the other hand, any birds you might attract under the best of circumstances are accustomed to the general level of human and animal activity usual in your area. If they weren't they wouldn't be in your area in the first place. So an extreme degree of sensitivity isn't to be expected from the birds, and need not be exercised by you. Furthermore, unless the birdhouse is placed very close to a window, the birds will pay little attention to what is going on inside the house.

As with the feeder, an excellent place for a birdhouse is one visible from the breakfast area. Birds are early risers, and it will be possible, each morning at breakfast, to see what is happening with the tenants. Keep a pair of field glasses next to the sugar bowl.

Sooner or later you are going to find a baby bird that has fallen out of the nest and is sitting on the ground, pitifully begging for food. The best thing you can do is

The houses in this martin colony are made of ceramics. Notice the ceramic predator guards on the strings.

post a subtle watch for cats, and otherwise leave it alone. There is a good chance that its parents will find it and care for it. If they don't, there probably isn't much you could have done in any case.

In *How to Watch Birds*, Roger Barton tells of a woman in California who had some measure of success feeding nestlings "a basic food of equal parts of finely sifted bread crumbs and finely mashed yolk of hard-boiled eggs," slightly moistened with milk and given with a medicine dropper. After about a week "she adds to this mixture some finely sifted corn meal for seed-eating birds and finely chopped meat or worms for insect eaters." Feeding is required at least every fifteen minutes from dawn to dusk, later once an hour. As the bird gets older, it is necessary to teach it to feed itself. Barton doesn't say how you do that.

What about putting the little bird back into the nest? Is it true that the mother will reject a nestling tainted with human touch? No one really knows. Sometimes the feeding parents will take up caring for the bird as though nothing had happened. On other occasions they will push the baby out of the nest, or even peck it to death. Not in-

WORKING OUT A DESIGN

It is fun to work out the design of your little house on paper, before actually building it, and not at all hard to do.

Basic to my own designing kit are plain white typing paper, a sharp pencil, a clear plastic drawing triangle, and an architect's scale—this is a ruler of triangular cross-section which has twelve different scales, allowing you to read off directly the "real" feet or inches of a scale drawing. A compass for drawing circles, a protractor for measuring angles, and a French curve for drawing curves of changing radius are sometimes handy. Some people prefer to use graph paper, in which case the architect's scale and drawing triangle can be dispensed with, but I like the architect's scale because it offers a wider choice of scales than graph paper generally does, and I like the look of a finished design on a plain paper. Graph paper, on the other hand, allows measurements to be read off directly, if you know the scale. If you use graph paper, be sure to get the plain kind, which has only small squares, or the kind with the big squares divided into 12, not 10, little squares.

It is a good rule always to make the drawing as large as you can on your paper. The larger the scale, the more easily you can show small detail, or small measurement differences.

Remember when drawing that building lumber is generally smaller than its nominal dimensions (see Dimensions of Boards, page 28), while building sheets (plywood, Masonite, etc.) really are the sizes they are called. But watch out for such as tongue-and-groove plywood flooring or shiplap boards. Because two or more pieces overlap in fitting together, their sum measurement will always be less than the total individual measurements of all the pieces, and that must be allowed for. Reckon shiplap or tongue-and-groove boards to be an inch narrower than nominal when figuring area to be covered, and deduct ¼ inch from the total width and total length of each tongue-and-groove plywood sheet.

frequently, it has happened that parents have abandoned the whole nestful of babies after someone has replaced one that has fallen.

BIRDHOUSE DESIGN

Birds are concerned exclusively with the amenities of their house, its neighborhood, and its convenience. Whether it looks like the Taj Mahal or a recycled cigar box makes no difference to the birds. In fact, it can *be* a recycled cigar box. You have virtually unlimited scope to exercise your imagination when it comes to designing a bird-house. Think how spiffy a tiny model of your house would look, sitting on a post outside its spiritual parent. If you have a small house, go in for wrens. If you have something of the order of Blenheim Palace, think about attracting a colony of purple martins or a pair of barn owls.

Do you like cars? Get a largish model car kit from a toy store, put it together in a tidy way, and make it into a birdhouse. But omit the seats and steering wheel, leaving the interior as uncluttered as possible. A closed car might well attract wrens or chickadees. If you build an open car, mount it on a sheltered spot and see if you don't collect a family of robins or swallows. Keep in mind, though, that the interior must be at least 6 inches square.

All the birdhouse-nesting birds are very good at grabbing and clinging, so there is no need to make the edges of the hole especially rough. Do not, however, take pains to smooth it, either. The birds do need a little something to get their claws into.

PROTECTING THE BIRD FEEDER

The same double-threat arrangement may be used below a bird feeder, although then it must be considerably closer to the ground, as you need to be able to clean both the bath and the feeder frequently as well as replenish the food supply. Protecting feeders is more difficult than protecting birdhouses, because a reasonably athletic and determined cat can leap as high as it is usually convenient to place a feeder. Still, the arrangement suggested will go a long way towards discouraging cats and squirrels and so forth. And consider this: If a cat manages to leap high enough to get his front paws hooked around the edge of the bath, the metal will bend and the cat will get a face full of water. This will tend to reduce his level of enthusiasm for the project.

CLEANING UP

Whatever birdhouse you build, it is advisable to devise a way to open it up, to allow you to get into the nesting compartment and clean it out. Birdhouses should be cleaned out once a year. Some experts on the subject recommend using a disinfectant, but unless you had some indication that the previous occupants were diseased, that doesn't seem necessary.

The best time to clean the birdhouses is soon after the nesting period is over and the birds have definitely left it for the year. That is also the time to put up a new birdhouse. The reason is simple: There is the whole of the winter for the smack of newness (or newly cleanedness) to wear off. Birds don't seem to like new or freshly cleaned houses as well as people do, so if you want the birdhouses to be occupied, you had best accommodate yourself to this eccentricity on the birds' part.

Shed-Roof Birdhouse

Here is a design for a basic, shed-roof birdhouse that can be cut from a single board leaving no waste. Look at the illustration, Fig. 1, page 29 to see how to cut the board.

Use a select grade of pine and inspect each board before you buy it to see that it is flat and straight, and that all knots are tight. There are cheaper grades of lumber, but cheap boards are seldom flat or straight, and low-grade fir splits more easily when you drive nails into it than does good pine.

The size should be suited to whatever species of bird you wish to have use it. Study the charts below for the size of the house, the diameter and placement of the hole, height from the ground to mount the birdhouse. Note that by eliminating the front wall, you have a roofed nesting shelf that will accommodate a family of phoebes, robins, sparrows, or swallows.

The lengths of the floor sections are left somewhat indefinite, as the exact dimension will depend on the planed width of your board. The number assumes your board will be ¾ inch thick and ½ inch narrower than its nominal width; it may be even smaller. Boards from different mills vary somewhat so do not cut out your floor piece until you are able to determine from the assembled side pieces just the size piece you must have for a snug fit.

The "length board" is also left indefinite, and you will find that a total length is called for that is greater than the sum of the several individual pieces. In part that is because you need to allow for the thickness of each of the five cuts (⅛ inch is allowed here, and is ample), and in part to permit some variation in the length of the floor piece. It might be a good idea to allow an extra inch for error.

SIZES OF BIRDHOUSES

ALL SIZES IN INCHES	FLOOR SIZE	INSIDE HEIGHT AT FRONT	SIZE BOARD	LENGTH BOARD (approx.)	SIDE WALL		FRONT WALL	BACK WALL	ROOF	FLOOR (approx.)
					FRONT EDGE	BACK EDGE				
small birdhouse	4x5½	8	1x6 (¾x5½)	50	8	9	8¾	10	8	5½
medium birdhouse	6x7½	12	1x8 (¾x7½)	71	12	13½	12¾	14½	10	7½
large birdhouse	8x9½	12	1x10 (¾x9½)	76	12	14	12¾	15	12	9½
nesting shelf	6x8	8	1x8 (¾x7½)	47	8	9½	none	10½	10	8

TAILORING THE BIRDHOUSE TO THE BIRD

To make your house most inviting, drill a hole of the right size with its bottom edge the right height above the floor, and mount the house at the height shown.

SMALL BIRDHOUSE	HEIGHT OF HOLE ABOVE FLOOR (INCHES)	DIAMETER OF HOLE (INCHES)	HEIGHT ABOVE GROUND (FEET)
Bluebird	6	1½	5–10
Chickadee	6	1⅛	6–15
Titmouse	6	1¼	6–15
Nuthatch	6	1¼	12–20
House & Bewick's wren	5	1⅛	6–10
Carolina wren	5	1¼	6–10
Violet-green & tree swallow	3	1½	10–15
Downy woodpecker	6	1¼	6–20
MEDIUM BIRDHOUSE			
House finch	4	2	8–12
Starling	10	2	10–25
Golden-fronted & redheaded woodpecker	10	2	12–20
Hairy woodpecker	10	1½	12–20
Saw-whet owl	9	2½	12–20
LARGE BIRDHOUSE			
Flicker	9	2½	6–20
Screech owl	9	3	10–30
Barn owl	4	6	12–18
Sparrow hawk	9	3	10–30
Wood duck	8	4	10–20
NESTING SHELF			
Robin			6–15
Barn swallow			8–12
Song sparrow			1–3
Phoebe			8–12

SHED-ROOF BIRDHOUSE

Materials Needed

one board, the size specified in
    chart

about twenty 2-inch finishing
    nails

carpenter's glue

two 1½-inch butt hinges with
    screws

hook and eye

exterior oil-base enamel paint
    (not red)

Construction Steps

1. Cut your board to the total length shown in Fig. 1. Make sure both ends are square.

2. Measure for each piece and cut it out before going on to the next. When you reach the cut between the front and back walls, tilt the saw about 12° from the vertical, so that the bottom of the cut is displaced into the back wall piece by about ³/₁₆-inch. That will give you the beveled tops you want for the front and back walls; the dimensions given refer to the longer face of the board. Cut the floor piece from the roof piece, but do not trim it to size.

3. Using an auger or a hole-cutter, make the entrance hole. Consult the chart on page 27 for the size and height preferred by the kind of bird you want to attract. Note, though, that this house is only suitable for small birds. If you don't have a particular species in mind, make your hole 1½ inches in diameter.

4. Assemble the walls as shown in Fig. 2, using glue and nails. Note that the front and rear walls overlap the floor, while the side walls do not.

5. After you've given the glue a chance to harden (follow the directions for the glue you're using), trim the floor piece to fit snugly between the front and back wall overlaps, and nail and glue it into place.

6. Attach the roof with the hinges. At the front attach the hook and eye. Alternatively, nail down the front of the roof, leaving ¼ inch of the nail protruding, so you can draw it out once a year for cleaning.

7. Paint the birdhouse with two coats of oil-base enamel.

DIMENSIONS OF BOARDS

Lumber can be bought either rough-cut or dressed—some say planed. Rough-cut lumber is *full dimension*, that is, a rough-cut 2x4 actually measures 2 inches by 4 inches. Dressed lumber has had its surfaces smoothed and some of the wood is lost in the process. Wood also shrinks as it dries.

What with dressing and drying, then, a board will be up to ½-inch narrower than its nominal width; a 1-inch-thick board will be about ¾ inch, and a 2-inch-thick board will be about 1½ inches. Remember that throughout this book we are talking about dressed lumber and not the rough-cut variety, so a reference to a "2x4" is a short way of saying "a board nominally 2 inches by 4 inches, and really about 1½ inches by 3½ inches."

Also keep in mind that the lumber you buy isn't likely to be precisely the size you expect. Since variations can be as much as ⅛ or ¼ inch, be sure to measure the lumber you get, and adjust the dimensions in the plan if necessary.

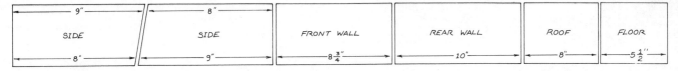

SIDE	SIDE	FRONT WALL	REAR WALL	ROOF	FLOOR

9″ / 8″ for first SIDE, 8″ / 9″ for second SIDE, 8 3/4″ for FRONT WALL, 10″ for REAR WALL, 8″ for ROOF, 5 1/2″ for FLOOR

Fig. 1

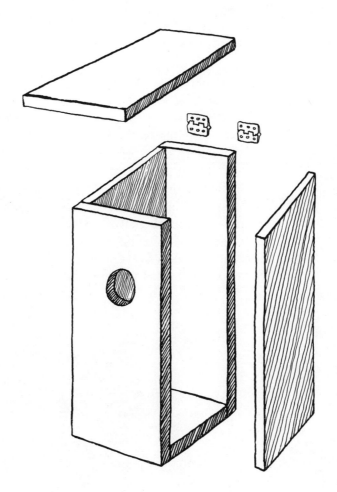

Fig. 2

Peak-roof Birdhouse

This simple birdhouse makes a fine home for songbirds. It is easy to assemble and can be mounted directly onto a post.

Materials Needed

about 4 square feet of ½-inch exterior plywood (all pieces can be cut from a strip 6 x 57 inches if you're careful)
three 4-inch common nails
box of 1-inch finishing nails
carpenter's glue
exterior oil-base enamel (not red)

Construction Steps

1. Cut out the eight pieces of plywood. Measure carefully and check the size of each piece before going on to cut the next. For dimensions, refer to Fig. 3.

2. Make an entrance following the directions under Shed-Roof Birdhouse, step 2.

3. Assemble the walls as shown in the drawing, using glue and 1-inch finishing nails set about 2 inches apart. The nails are mainly to hold the glued joints tight until the glue dries.

4. Glue and nail the roof halves together and to the walls.

5. Check to see that the 5-inch-square base piece fits easily inside the base of the walls, and that when you place the four walls down on the larger 6-inch-square base piece, the outside of the walls is flush with the edges of the base piece. If necessary, trim to fit, then glue the two base pieces together; the smaller one centered on the larger one. Do not fasten the base to the walls yet.

6. Paint everything with two coats of oil-base enamel.

7. When the paint is dry, nail the base unit to the top of the post (see page 38, "How to Sink a Post"), using three 4-inch common nails. Angle each nail slightly towards the middle of the post.

8. Fasten the walls and roof of the house to the base with two nails, driven horizontally through the very bottom of the walls into the upper base piece. Leave about an eighth of an inch of each nail proud of the surface, so you can open the house for cleaning by drawing out these nails.

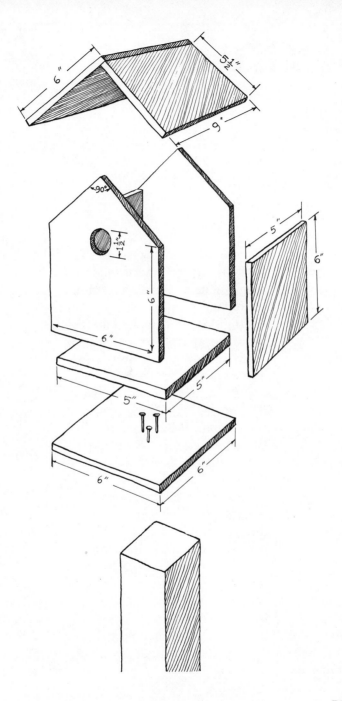

Fig. 3

Traditional Martin House

Martins are very satisfying birds to build a house for. For one thing, being plentiful, and distributed over the whole of the United States and virtually all of the settled areas of Canada, they are relatively easy to attract. They arrive from South America, where they winter, as early as the end of January in Florida, and by early May they are as far north as Manitoba. They are amenable to either rural or semi-urban environments. For another thing, they are pretty little birds—members of the swallow family—and you can get a lot of birds for your birdhouse dollar.

Martins houses can take quite a few different forms. Basically there are two sorts—single- and multiple-unit models. The traditional martin house has six or more cubicles (sometimes many more, even hundreds) in a single house. I have seen them round and square, spread out like a bungalow and stacked up like high-rise apartments. the birds seem to show no preference for one style over another.

The martin house on pages 34 and 35, Figs. 5 and 6, is traditional. It is square in plan, has a peaked roof, and it will accommodate 24 pairs of nesting martins. The cubicles are 6x6x6 inches. Some martin fans recommend "apartments" as large as 8x8x8 inches, but I prefer to stay with the smaller size. It has the advantage of making for a lighter birdhouse that is more compact and offers less wind resistance. It is also easier to mount on a post than houses that are made up of larger cubicles.

TRADITIONAL MARTIN HOUSE

Materials Needed
one and a half 4x8-foot sheets of $^3/_{16}$ inch plywood, with one good side

$^3/_{16}$-inch-diameter threaded rod, 27 inches long

2 flat washers, 1 ordinary nut, and 1 wing nut for the threaded rod

box of ¾-inch brads

carpenter's glue

four 4-inch common nails

Special tools needed
To saw the plywood, use a table or radial-arm saw, although if you are careful to stay right on your cut lines, you can use a portable contractor's saw or an ordinary hand saw. In all cases, use a blade with fine teeth—often called a "plywood blade."

Construction Steps
1. Cut out all the pieces as in the cutting diagram, Fig. 4

A. 3 floor pieces, 22¾ inches square

B. 4 exterior wall pieces, 6 inches x 18 inches

C. 6 exterior wall pieces, 6 inches x 17⅝ inches

D. 12 interior wall pieces, 6 inches x 17⅝ inches

E. 2 gabled exterior wall pieces, 18 inches long, 6$^3/_{16}$ inches high

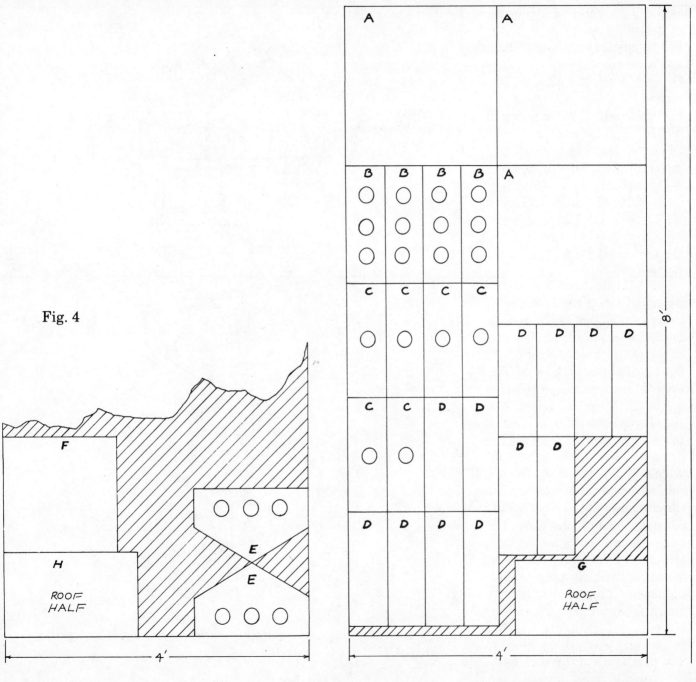

Fig. 4

ROOF
HALF

ROOF
HALF

at the ends, and 11½ inches high at the gable peaks

F. 1 third-story ceiling, 18 inches x 17⅝ inches

G. 1 roof half, 12 inches x 21 inches

H. 1 roof half, 11³/₁₆ inches x 21 inches

24 corner locators, ½ inch x 1 inch (not pictured; see step 5)

    2. Drill one 2¼-inch hole in each of the 17⅝-inch exterior wall pieces (C). The center of the hole should be equidistant from the ends, and 3½ inches from the bottom.

    3. Drill three 2¼-inch holes in each of the 18 inch exterior wall pieces (B) and the gabled wall pieces (E). The center of the center hole should be equidistant from the ends. The other holes should be centered 3³/₁₆ inches from each end. All three holes should be 3½ inches from the bottom of the wall.

    4. Cut two slots, ³/₁₆ inch wide and 3 inches deep, in each of the interior wall pieces (D). The inner edges of the slots should be 5¾ inches apart, and the outer edges should be 5¾ inches from the ends of the piece. The easiest way is to gang-cut them, that is, cut them all together in two operations, one for each slot in all the sections. If you use a table or radial-arm saw, set the blade to saw at a depth of three inches

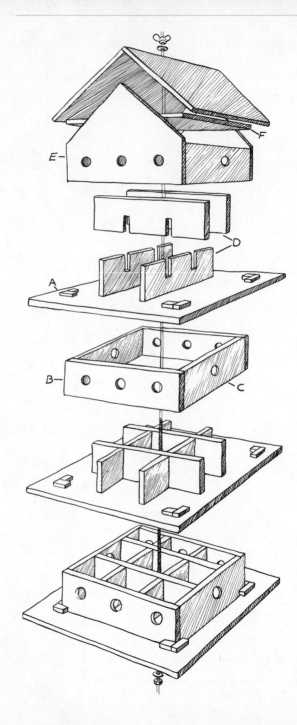

Fig. 5

from the table top. Place all the pieces together, resting on their long edges. Keep the ends even by holding them between two short pieces of board and clamp with a C-clamp. Several passes will be necessary to get a width of ³⁄₁₆ inch. If you use a contractor's saw or a hand saw, clamp the pieces together in a vise before sawing the slots. Be sure to pad the vise jaws with pieces of scrap to avoid marring the wall pieces.

5. Make the 24 corner locators by cutting a ¼-inch strip of the plywood, 24 inches long, and cutting it into 24 equal pieces.

6. Glue and nail the wall pieces together, two shorter C pieces and two longer B pieces per "floor." Use 4 brads for each joint. Clamp the joints, or set the "boxes" on end, three-hole sides top and bottom, and weight the ends with something heavy. But be sure to brace them on the sides so the walls stay square. A piece of scrap laid diagonally across the walls and temporarily nailed in place will do the job. Assemble the top wall the same way, using pieces C and E.

7. Place one finished wall square on each of the floor pieces, centered so that there is a 2⅜-inch margin all around. Mark the corners with a pencil.

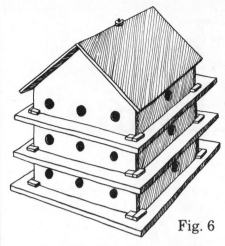
Fig. 6

8. Glue the corner locators in place, outside the lines you have just drawn, two to each corner. Clamp them in place by weighting while the glue dries.

9. Drill a ³⁄₁₆-inch hole in the exact center of the third-story ceiling. Glue and nail the ceiling onto the side walls, and against the insides of the gabled walls.

10. Glue and nail the roof halves to the gabled walls. Use a brad every 2 inches. Do not try to nail the seam where the two roof halves join until the glue has thoroughly set.

11. Drill ³⁄₁₆-inch holes in the exact center of each of the floor pieces, and in the center of the roof peak.

12. Slot the interior walls together into three units of four pieces each. No gluing is necessary.

13. Put one washer on the threaded rod and screw the plain nut on until its bottom is just flush with the bottom of the rod. Poke the rod through the hole in the bottom floor section.

14. Put a wall section in place, fit an interior wall unit into it, and lay the second floor section on top.

15. Assemble the rest of the house in the same way. At the top, place the remaining washer over the end of the threaded rod, and fasten the whole structure together by screwing down the wing nut. Don't screw it so tightly, though, that you split the roof or the bottom floor section.

16. Paint the house with an oil-base exterior paint. Two coats should be enough for the flat surfaces, but the exposed plywood edges may well need three or four coats. You may disassemble the floors of the house for painting, but do not paint until all the gluing is done. Paint on surfaces to be glued will weaken the joints.

17. To mount this house on a post, you must first make a small cavity in the top of the post to accommodation the nut. Then, using four 4-inch common nails, nail the bottom floor section with the threaded rod in place to the post top. Then assemble the house.

Multi-Tube Martin House

Here is a design for a less traditional martin house; one made of sections of the hard cardboard tube around which new rugs come wrapped. Rugs used to come rolled up around sticks of bamboo, which was nice because bamboo is a very handy material for all kinds of things, from walking canes to the very best kite sticks. But nowadays rugs usually come rolled up around long, very stiff, cardboard tubes, ranging in inside diameter from 3 to 8 inches. For a martin house you need at least a 6-inch tube and larger if you can get it.

MULTI-TUBE MARTIN HOUSE

This is a particularly simple-to-build martin house for six families, see Fig. 7.

Materials Needed

heavy cardboard tube, inside diameter 6 to 8 inches, 4 feet long

two pieces of ½-inch plywood, each 8½ inches by 3¾ times the outside diameter of the tube

an equilateral triangle of the same plywood, each side as long as 3¾ times the outside diameter of the tube

1-inch board no narrower than the inside diameter of the tube and as long as six times the inside diameter of the tube

weatherproof wood glue

three 1-inch #8 flathead wood screws

four ¼-inch pan-head screws

exterior paint

four 4-inch common nails

Construction Steps

1. Cut the tube into six equal 8-inch segments.

2. Smooth all the edges with sandpaper.

3. Paint the tube segments. An easy way to do this is to poke a little hole into one side of each tube with an icepick or something similar and thread the hole with a foot-long length of string or wire. You can then simply dip the tubes into your paint and hang them on a clothesline or tree branch to dry. If your paint can isn't deep enough to dip a whole segment, dip the end with the hole first, before you thread the wire, then turn it over, thread the wire, and dip the other end.

4. Cut your plywood to the dimensions indicated above, and sand the edges smooth.

5. Glue and, using the 1-inch screws, screw the triangle to one of the long sides of one of the plywood rectangles, with the bottom of the triangle resting on the rectangle.

6. Paint the rectangle-triangle assembly, except for the bottom of the rectangle. Also paint the edges and one side of the other rectangle.

7. In a triangular stack, glue the tube segments to the vertical triangle, and to the rectangular base. Also glue them to one another.

8. While the glue is setting, cut six disks from the 1-inch board, each disk to be a tight fit into the end of a tube segment. If you don't have a hole cutter the

right size, cut squares of the same dimension as the inside diameter of the tubes, and by progressively cutting off points, dress the square into a circle. Finish with a rasp and sand paper, but be careful not to get the disks too small.

9. In the center of each disk, cut a hole 2½ inches in diameter.

Leave rough. The birds like a rough entrance.

10. Paint the disks. These, too, can be dipped and hung with wire to dry.

11. When the paint is dry, fit the disks into the open tube ends. They should be a tight fit. If necessary, wrap a loose disk with a turn of black friction tape.

12. Nail the second rectangle, unpainted side up, to the top of a properly set post (see page 38). Then using the four pan-head screws, one at each corner, screw the house base onto the post platform, and wait for your martins.

Fig. 7

Mounting a Birdhouse

Unless you have a good reason to want to suspend your birdhouse from a wire, I recommend you mount it to a tree or post. The post is preferable for two reasons. First, if the house is post-mounted, it is easier to protect the birds from predators. Cats and raccoons, for example, find it easy to climb a tree and walk out on a limb to loot a birdhouse. Even if you protect the birdhouse by mounting a predator guard on the limb, it may still be possible for some predators—cats in particular—to drop down from another branch, or leap up from a lower one. A post, on the other hand, can have a good predator guard. And it may be so constructed that it can be lowered in order to make the birdhouse accessible for annual cleaning.

HOW TO SINK A POST

If your your birdhouse is a small one and you don't feel you need a post you have to lower, a plain, painted (and, if possible, rot- and insect-resistant) 4x4 is perfectly satisfactory. A hole in the ground 8 inches and 18 inches deep is large enough. Stand the post in the hole, on top of a double handful of coarse gravel or small stones, and temporarily brace it well in two directions so that it is vertical, and will remain so as you pour the hole full of concrete. Screed the concrete with a hoe handle or the like (see Step 7, page 72), and be sure it is mounded up somewhat around the post for drainage. Your job will be somewhat eased if you mount the birdhouse onto the post before you erect it. Don't remove the temporary braces until the concrete has well and truly dried—at least 24 hours. Annual cleaning will have to be done from a ladder, of course, but otherwise the rigidly mounted post should be as satisfactory as the tipping one below.

HINGED BIRDHOUSE POST

Here is a design for a post arrangement that makes it relatively easy to get the birdhouse when cleaning time comes around, see Fig. 8. When it is time to clean out the nests, you need only to remove either one of the two bolts and gently let the post with the house tip over to the side.

Materials Needed
one 4x4 post, the length specified in the chart on page 27
two 2x6 boards, 41 inches long
two 8-inch-long ½-inch bolts with two nuts and four flat washers to fit
one cubic foot of concrete
⅓ cubic foot of gravel
oil-base enamel paint

Construction Steps
   1. After cutting them to size, paint the 4x4 and the 2x6s with two coats of exterior oil-base enamel paint.

   2. Dig a hole 10 inches square and 2 feet deep; if you

wish, rent a posthole digger for the job.

3. Fill the hole to a depth of 6 inches with gravel or small stones.

4. Temporarily nail the 4x4 and the 2x6s together so that the 4x4 is sandwiched between the 2x6s for a distance of 21 inches. The 2x6s will extend beyond the bottom of the 4x4 by 20 inches. Leave enough of each nail protruding from the boards so that you can easily pull them out later with a claw hammer.

5. Drill holes ½-inch in diameter through the 2x6s and the 4x4. The top hole should be 1½ inch from the top of the 2x6s, and the bottom hole should be 1½ from the end of the 4x4.

6. Insert the bolts, with one flat washer under the bolt head. Put on the other washer, screw on the nuts, and draw them up tight.

7. Draw out the the nails temporarily holding the boards together.

8. Mix up one cubic foot of concrete. A metal wheelbarrow makes a good mixing trough; a hoe a good mixer. Wash them off thoroughly when you're finished. Directions for mixing concrete appear on page 98 of the chapter on privies.

9. Set the 2x6s, still bolted to the post, into the hole, resting on the stones or gravel. Check that they extend 18 inches into the ground—no more. If necessary, add or remove a little gravel. There must be at least 2 inches clearance between ground level and the bottom of the 4x4 post.

10. Brace the post in a vertical position (check with a level) using two boards, each at least 8 feet long, temporarily nailed to the post at right angles to one another, and at about a 45° angle to the ground. Recheck that the post is vertical.

11. Pour the concrete into the hole. With the handle of your hoe work the concrete in between the two boards, then poke it vertically, repeatedly and deeply, with the hole handle to work out any air bubbles. With a trowel or a piece of flat board smooth the concrete so that the outside edges are just at ground level, and the center, by the post, is about ½-inch above ground level. Recheck the post for verticality. When concrete is thoroughly dry, remove the braces.

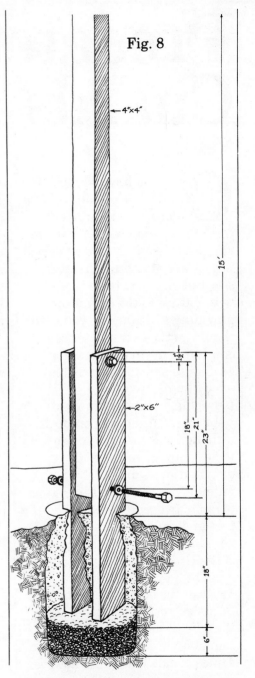

Fig. 8

Predator Guards

Once you have set up a birdhouse, you have some obligation to protect the occupants from predators. To be sure, they would not be protected if they were to nest in a natural site, but once you have interfered with nature to the extent of providing a birdhouse, you really should go the extra distance and do what you can to make the birdhouse safe for birds. There aren't many creatures, particularly around human habitations, that prey on birds, although the ones that do so tend to be common and of wide distribution. Quite a few animals, raccoons and opossums to name two, can climb a tree or a post and make a meal of nestlings or eggs. In some parts of the United States, snakes must be added to the list.

But the commonest and the most destructive of predators is the *felis domestica*, the common-garden-variety domestic cat. There are, however, ways to make bird-ravaging difficult for them. Here are two effective predator guards.

The first one consists of a sheet of aluminum wrapped around the post directly below the birdhouse, forming a slippery sheath that no creature will be able to climb past, Figs. 9 and 10. The second consists of a slightly dished circle of sheet aluminum with a hole cut out in the center for the post, see pages 42 and 43, Figs. 11 and 12.

SHEATH-STYLE PREDATOR GUARD

Materials Needed
rectangle of hardware-store
    sheet aluminum, at least 24
    inches on one side and, on the
    other, 1¼ times the
    circumference of your post
    plus one inch; .020-gauge
    aluminum will do nicely, but
    the thickness is not crucial,
    Fig. 9.
about twenty large-headed,
    galvanized 1-inch roofing
    nails

Construction Steps

1. Measure the circumference of your post. As noted above, the aluminum should be 1¼ times the circumference of the post, plus 1 inch.

2. Cut the aluminum sheet to size. If you don't have tin snips, you can probably have it cut where you buy it.

3. Nail one 24-inch side to the middle of one side of the post, directly under the birdhouse, with the bottom of the aluminum at least 6 feet from the ground, see Fig. 10. Start with a middle nail and pull the metal around to make sure you're not going at an angle.

4. Wrap the aluminum as tightly as possible, then nail the other 24-inch edge in place, and put a nail each into the top and bottom of each of the other sides.

5. With a hammer beat down all the exposed edges of the aluminum so that they won't cut someone.

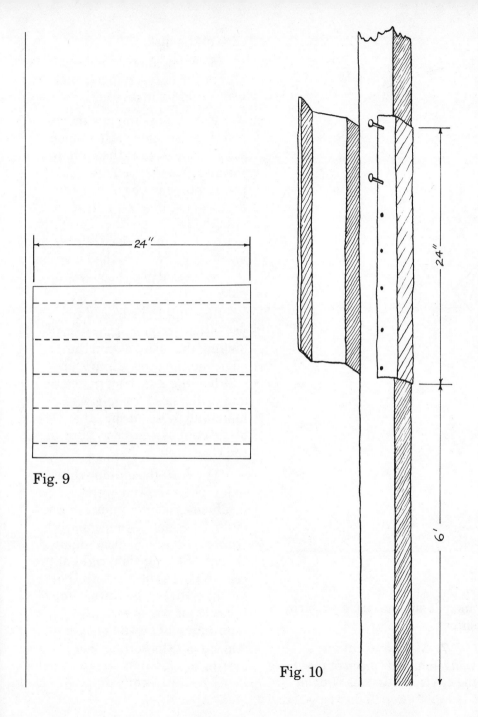

24"

Fig. 9

24"

6'

Fig. 10

DISH-SHAPED PREDATOR GUARD

Materials Needed

3x3-foot sheet of aluminum, .020 gauge

scrap wood about 3 inches square

20-inch piece of wire or string, plus two nails to form a "compass"

eight 2-inch finishing nails

pop rivets

caulking

Construction Steps

1. Cut out a 30-inch square piece of sheet aluminum with tin snips.

2. Find the center of the piece by lightly scribing diagonal lines corner to corner. Where they cross is the center.

3. Place the aluminum on the piece of scrap wood; drive a nail at the center mark through the aluminum and into the wood. Leave part of the nail sticking up so you can do step 4.

4. Cut a piece of thin wire or nonstretchy string about 20 inches long. Tie or twist one end around the center nail; leave it loose enough so it can rotate freely. Wrap the other end tightly around another nail, one long enough to get a good grip on, and continue wrapping until the free nail is exactly 15 inches from the

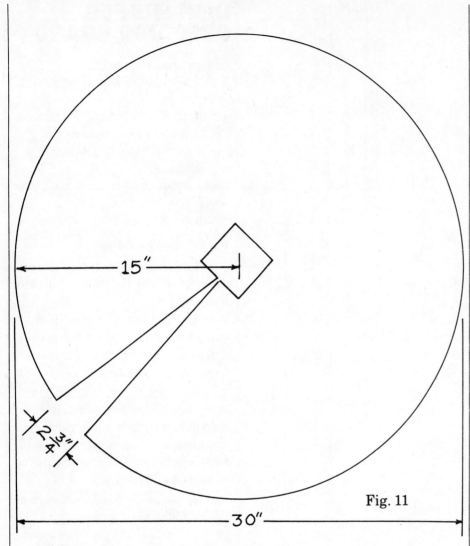

15″

2¾″
3″

30″

Fig. 11

edge of the disk.

8. From another piece of aluminum cut a strip 2 inches wide and 13½ inches long.

9. Overlapping one inch, pop-rivet one side of the 2-inch strip to one edge of the cut-out in the disk. An inexpensive pop-riveter can be bought at a hardware store. You'll have to hold the parts together carefully (with a C-clamp, Vise-Grip, or vise) while you drill the holes for the rivets and pop the rivets in place.

10. Pull the edges of the wedge opening together (thus dishing the disk), clamp the free edge and the protruding inch width of the 2-inch strip together, and drill holes for pop-rivets matching those on the other side of the cut-out—but don't put in the rivets yet.

11. With the clamp still in place, place the end of the birdhouse post (or a piece of wood cut to the same dimensions as a cross-section of the post) against the inside of the dish, right at the center where you want the hole for the post. Scribe the outline of the post (or pattern) on the aluminum. Remove the clamp and cut out the square. For this cutting job you may find it easier to use a sturdy knife or a keyhole

center nail.

5. Use the sharp point of the free nail to scribe a 30-inch circle.

6. Pull out the center nail and cut out the circle with tin snips.

7. Mark and cut out a wedge-shaped "piece of pie" from the circle, 2¾ inches wide at the hacksaw rather than tin snips, which will be difficult to position.

12. Partially drive four 2-inch finishing nails into the post, at the height at which the guard is to rest—at least 6 feet and preferably 8 feet above the ground. Leave about an inch of each nail protruding, and tap them down a little so that the contour of the inverted dish will rest evenly on them.

13. If the top of the post is free, finish riveting the guard together and slide it, dished side down, down the post from the top until it rests on the four nails. If the post is in the ground and there is an unremovable birdhouse on the top, you will have to work the guard on by way of the open wedge cut and do your best to straighten out the metal once it's around, before putting in the second row of rivets. In this case put the guard on at a level where it's comfortable to work, rivet it together there, slide it up high enough to drive in the nails *under* it, then drop it into position.

14. Drive four more finishing nails over the guard, but this time allow only a half inch or less to protrude. Tap the nailheads down until the guard is firmly held in place.

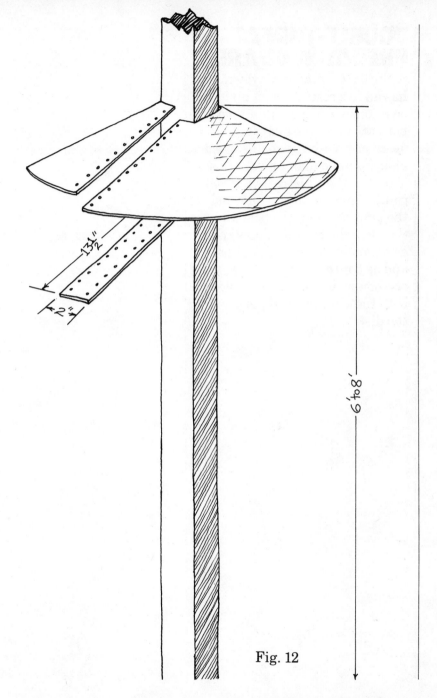

Fig. 12

DOUBLE-THREAT PREDATOR GUARD

A third anti-predator system that has an interesting twist all its own consists of a combination of both the predator guards described, with the dish-shaped disk mounted dish upwards, as shown in the illustration. If you caulk the seam in the disk and the joint between the post and the disk, the disk will serve both as a part of the predator guard and as a birdbath. No cat or raccoon will be able to get past both the aluminum sheath and the disk.

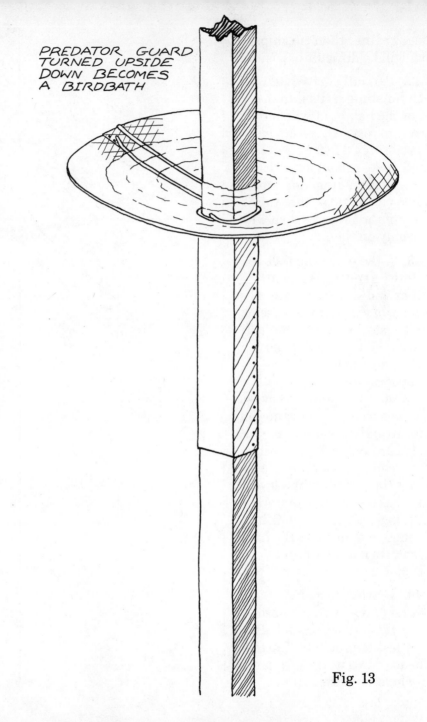

PREDATOR GUARD TURNED UPSIDE DOWN BECOMES A BIRDBATH

Fig. 13

DOGHOUSES

Should You Build a Doghouse?

Most dogs I know are as enthusiastic about living in a doghouse as they are about thunderstorms or toothache, but there are good reasons why a family might need or want to keep the dog outside. An outdoors dog will be less bothersome to a member of the family who doesn't like dogs or who is allergic to them. Training the dog not to wet the carpets or chew on table legs becomes unnecessary, dog hairs will not become a feature of every surface in the house, and the likelihood of fleas being introduced into the rugs and furniture is greatly reduced. Then, too, working dogs such as gundogs and watchdogs generally work better if they have been kept outside.

Years ago *The Saturday Evening Post* carried a cover picture of a freckle-faced little boy hammering together a charmingly incompetent doghouse. Off to the side a round-bellied puppy looks adoringly on, little suspecting that he is going to be asked to live in that instant slum, isolated from all he holds dear, which is to say, people in general and the freckle-faced little boy in particular. The boy, of course, is sure he is doing his dog a great big favor. A favor? To give his comfort-loving puppy a leaky, drafty, furnitureless, doorless, hard-floored hut to live in?

That is not to say that all doghouses are bad. Lots of dogs live in them all year round. A dog who has had a chance to acclimate himself to living outdoors, by doing so during the summer and the fall, will grow a progressively heavier coat as the weather gets colder, which, together with certain subtle changes to his circulatory and nervous systems, will make him uncomfortable in a heated house.

CAN YOUR DOG LIVE OUTDOORS?

Not all dogs, to be sure, can be kept outdoors. Very small dogs have a problem because their ratio of skin area to body volume

---

READY-MADE DOGHOUSES

It is entirely possible to "finish" a ready-made doghouse. The house must be elevated at least 2 or 3 inches from the ground. A door of some kind must be fitted to the opening. Arrangements for summer ventilation must be made, and there must be a way to get into the house well enough to clean it out thoroughly. These last two requirements can probably best be achieved by removing and hinging one or both halves of the roof, as is described in the plan in this chapter. Cracks in the floor, walls, and roof need to be caulked, and, the inside should be lined with indoor-outdoor carpeting.

is higher than that of larger dogs, and their body heat is thus more easily radiated off. If they are also short-haired, the problem is compounded. Chihuahuas and small terriers, for example, must remain house dogs, unless you live in a place with very warm winters.

Long-haired small dogs, like cockers or (unclipped) toy poodles, can stand a fair amount of cold, as can most short-haired large dogs, but they may not be able to remain warm enough in severe or prolonged cold weather.

Best able to take the cold are large, long-haired dogs, such as St. Bernards, Newfoundlands, samoyads, Norwegian elkhounds, huskies, and the like. Many individuals of these breeds continue to prefer life outdoors even if they are brought up as house dogs.

No dog, especially no puppy, should be thrust outdoors to live after he is accustomed to lots of warmth and company. So, if you buy a puppy and want to keep him outside from the start, get him in the spring, after you have stopped having cold nights. Provide him with plenty of bedding, particularly for the first few weeks. During the summer and fall, as he grows, he will have time to accustom himself

FAMILY LIVING

If your dog has puppies, you have another set of problems to consider. During nursing, of course, they must stay with their mother, and there will not likely be enough room in the doghouse for the whole family. Set aside an area in your basement or garage for that period. After they are weaned, if the weather is warm, or when they are about four months old, if it is winter, they can go outdoors. But in the latter case the transition must be a gradual one, to allow them to acclimate. At four months, four or five pups can occupy one doghouse designed for an adult of the same breed; any extra warmth they may need they will get by cuddling together. As they get older, of course, other, more permanent arrangements will have to be made.

Still, there is no reason why a doghouse should not be built to house more than one dog, providing the two (or more) prospective occupants get along well, and provided that one dog is not finally asked to occupy alone the cavernous space originally planned for three.

gradually to harsher weather. If he howls or whines at night, try giving him a ticking clock or a transistor radio turned low for company.

Unhealthy dogs and dogs beginning to suffer from old age should, of course, receive special consideration. Arthritic conditions are aggravated by cold and damp, and impaired circulation also means, just as it does in old people, impaired ability to stay warm. Special consideration, however, does not necessarily mean bringing the

dog inside. The dog, particularly if he is a large, long-haired sort, may well do best in a good, snug doghouse.

If you keep your dog outside during the winter, don't comb him out. Let his hair get long and matted; it's what's keeping him warm. If you comb out all the loose hair, you will seriously impair the ability of his coat to retain body heat. If he is a long-haired dog, he may get to look a bit scruffy, but love him as he is. It's natural, and all for the best.

DOGS' TASTES AND NEEDS

If you think your dog's tastes for creature comforts are less developed than your own, you are deluded. The dog has a finely honed appreciation of the cushy. I have often seen a dog rise from a nicely carpeted floor, walk across the room, and plunk down again on top of a handkerchief that has slipped off the ironing board. The rug is better than the bare floor, but a rug with a handkerchief on it is better yet, and the dog is no fool. The princess who complained about the pea under twenty mattresses differed in sensitivity from your average dog only in that she complained.

So if circumstances require that your dog live in a doghouse, two key words apply: soft and snug. Aesthetics play no role as far as the dog is concerned; as long as the place is soft and snug, it can look like a pile of used lumber or St. Basil's Cathedral.

FITTING THE DOGHOUSE TO THE DOG

In designing a doghouse, the first thing you must do is to figure out the size house your dog will need.

Start with a burlap bag.

This late 18th-century doghouse, complete with the royal arms on the gable, was built for an occupant named "O. Wadcock," and may well have been a replica of his master's house. It is commendably weather-tight, and the holes over the side door indicate that thought was given to ventilation. It looks as though O. Wadcock had to wait for someone to let him in and out, however.

CHOOSING A LOCATION

For the most part, the site for your doghouse can be dictated by convenience. If possible, put the house near (or better yet under) a good shade tree. Or on the north side of a building large enough to provide shade. A good doghouse can protect an acclimated dog from winter cold, but summer heat is something else again. Even a well-ventilated doghouse gets hot during the middle of a sultry summer day, and if the dog has to choose between a hot doghouse and the direct rays of the sun, he has a Hobson's choice indeed. Shade is important.

Then, if you can, locate the house on a small rise, or on a piece of ground that drains well. Not that water (short of a flood) will run into a properly elevated doghouse, but if the ground remains wet and muddy after a rain, much of that wetness and mud will get tracked into its house by the dog and cause discomfort.

This, when filled with cedar shavings, as described below, will completely cover the floor of the doghouse and will serve as the dog's bed. Take your other dimensions from it. If you have a large dog, a standard burlap bag, about 2x3 feet, should be about the right size. Feed stores, hardware stores, and lumberyards generally have them, or can help you get one. To check for size, lay the bag out on the floor and encourage your dog to lie down on it. This should not be difficult, as the bag on the floor will be slightly softer than the floor without the bag, and the dog will recognize this fact at once.

When the dog is lying on the bag in his favorite sleeping position, see that there is enough room on all sides, allowing for some shrinkage in area when the bag is stuffed. Keep in mind that too small is no good, and so is too large. Dogs prefer cozy to spacious. If your dog likes to circle a few times before lying down (most dogs do), make sure there is just enough room for the dog to do his four-legged pirouette.

When you have the right-size bag, fill it with cedar shavings until it is soft and pillowy. (Cedar shavings are available in any pet-supply store.) Don't overstuff the

Victor H. Lane

This doghouse is made of a few boards, a sheet of plywood, and an old packing case lined with rags. It is too wide open to suit most dogs I know, although it is placed in a sheltered doorway, and there is an electric light that the occupant's master says is more for company than warmth. The packing case slides out for easy cleaning, and can be taken inside as a dog bed on really cold or wet nights.

bag or it will get hard and lumpy.

Two remaining dimensions that must be tailored to your dog are the interior height of the walls and the size of the door opening. Ideally your dog should have just enough height inside the doghouse to do his circling. That, of course, doesn't mean full-standing height. Watch your dog when he circles to lie down. He does it in a sort of half-crouch. Take his height in that position as an average inside height for the roof—that is, the inside height halfway between the eaves and the peak.

The door opening should be just big enough for the dog to pass through without squeezing. The smaller the opening can be made and still allow the dog to pass in and out, the less likelihood there is of drafts and the cozier the house. To find out how big the opening must be, get your dog to lie down again, this time on his stomach in a sphinxlike posture. Run a tape measure around him at the shoulders, including his elbows. For our Labrador retriever that measurement is 33 inches. To find the diameter, divide by 3.14 (pi), and then add an inch and a half to allow the dog to enter comfortably; 12 inches is the diameter of the opening I would have to cut for our dog (see step 9, page 54).

Building a Better Doghouse

Most doghouses are sorry affairs, designed with little thought of the dog's well-being. In contrast, you will find here an uncommonly spiffy doghouse of my own design that is soft and snug, warm in winter and cool in summer. The floor is padded and elevated from the ground, the opening is equipped with a weathertight swinging door that will keep out drafts in winter and that can be hooked open for the summer. The house can be insulated for added warmth. The halves of the roof can be raised partway for ventilation and opened completely for cleaning. The shape of this doghouse is solidly conventional, but there is no reason why the house cannot be ornamented and decorated in ways to suit the owner's taste, see Fig. 14.

Materials Needed

Obviously, the amount of materials needed will vary depending on the size of your dog. Using the measurements you found above, you can figure out the dimensions for the pieces yourself. Below is a general list of what you will need to build a better doghouse.

¾-inch plywood (for the floor)
½-inch plywood (for the walls)
¼-inch plywood or hardboard (for door)
2x4s (for skids)
2x2s (for studs and shoes)
rubber strips from a truck or tractor inner tube
staples or ½-inch roofing nails
four strips of lath—2 as long as the house, 2 short
two 2-inch butt hinges
four 3-inch hooks and 8 eyes
one 6-inch hook and eye
1¾ nails (for nailing siding to studs, floor to skids)
1¾ inch siding nails
¾-inch box nails
3-inch (10d) common nails (for nailing shoes to studs)
carpenter's glue (optional)
indoor-outdoor carpet (optional)

Note: Half-inch plywood is called for for the side walls. This is perfectly adequate for both strength and insulation, as further insulating material can more easily be added to the walls inside than it can be to the floor (see box on page 57). But, if you are building a fairly small doghouse and could cut all floor, wall, and roof panels from one sheet, then by all means use ¾-inch plywood for the walls as well as for the floor and roof. You can also scale down a small doghouse considerably by doing away with the middle wall studs—but not the corner studs. And you can use 1x2s instead of 2x2s for the shoes. Don't go overboard in using lighter lumber, though. The house needs weight for stability (those eaves are great windcatchers), and light, thin lumber warps easily.

Construction Steps

1. Saw the floor out of a sheet of ¾-inch plywood.
2. Nail the floor onto the

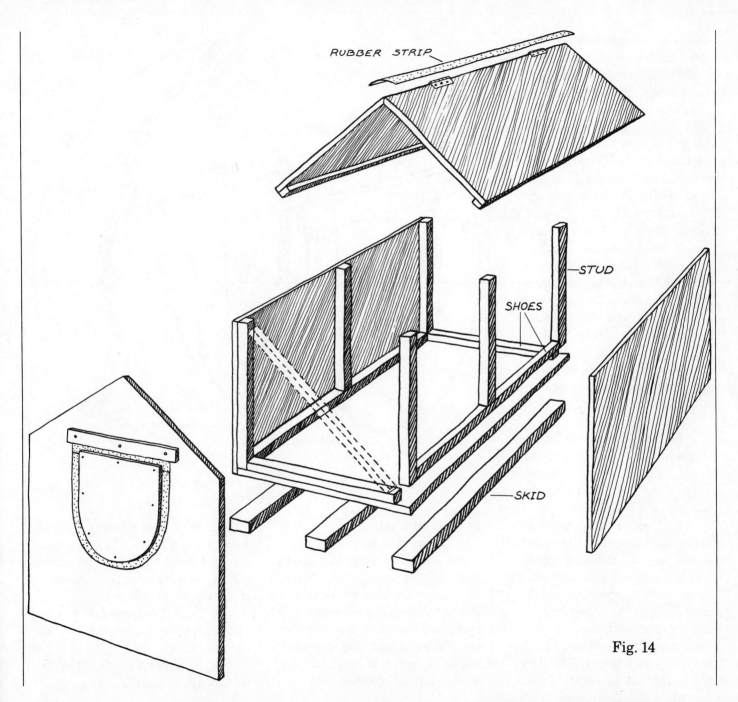

RUBBER STRIP

STUD

SHOES

SKID

Fig. 14

three 2x4 skids; these will form the foundation of the house. The two outside skids should protrude ½ inch beyond the side edges of the floor to provide a base for the side walls. All three skids should extend ½ inch beyond both front and back ends of the floor, in order to support the front and back walls. (If you are using ¾-inch plywood for the walls, the skids should protrude ¾ inches.)

3. Nail the center 2x2 side-wall studs (which are shorter than the corner studs) to the center of the 2x2 side shoes. Nail up through the shoes into the ends of the studs.

4. Make sure the four shoes are the right length; when nailed in place, they should just leave room for the four 2x2 corner studs.

5. Nail the four 2x2 corner studs in place, butted against the shoes, and flush with the corners of the floor. Nail horizontally into the butts of both shoes.

6. Now, using a length of scrap lumber, brace the side-wall studs, one at a time, by wedging the scrap at one end against the shoes on the opposite side, and holding it at the other end against the stud, near the top, as shown in Fig. 14. When the stud is steady, nail the side wall in place. Repeat for each

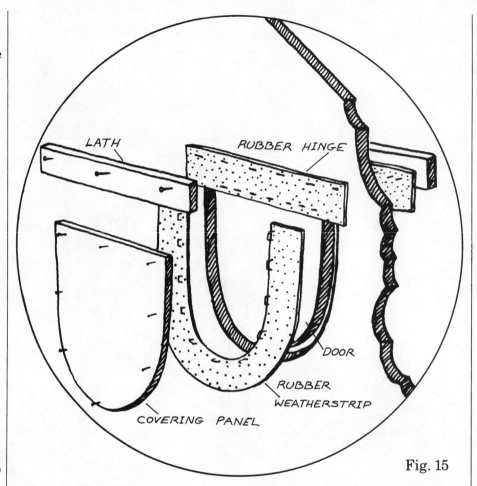

Fig. 15

remaining stud.

7. You now have the foundation, the floor, and the two side walls in place, and it is time to work on the end walls. Cut out the two identical end walls. The roof pitch shown in the drawing is not critical (it has 6 inches of rise to 14 inches of run), but it strikes a good compromise

between the heat conservation of a low ceiling and sufficient pitch for good drainage. The end walls should be wide enough to cover the ends of the side walls.

8. Nail the back wall in place. Its base, too, should rest on the skids.

9. Now, before attaching the front wall, make the swinging

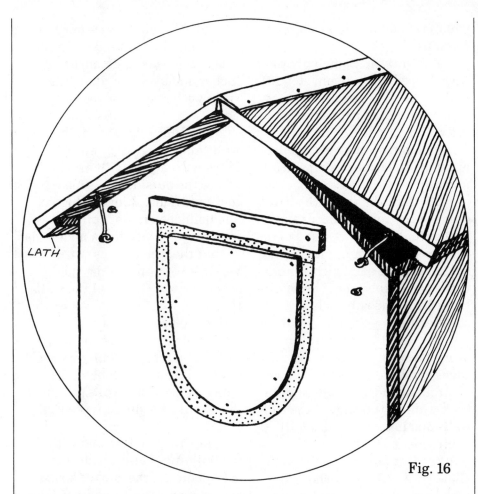

LATH

Fig. 16

perfectly rectangular piece of wood or stiff cardboard), square off the top of both circles, so you have one ⊔-shape within another, Fig. 15.

12. Cut out the opening with a jig or saber saw. If you don't have one, rent one or borrow one for the job. Start the cut by drilling a hole in the 1-inch margin. Be especially careful in your cutting! It is difficult to file or sand a raw plywood edge without tearing up the outer layers of wood, and you will need the cutout piece for the door. Smooth the opening carefully with a file or sandpaper, rounding the sharp edges.

13. Smooth the edges of the ⊔-shaped door with sandpaper. Trace its shape on a piece of ¼-inch plywood or hardboard, then cut the thinner door shape out and smooth its edges. Save the thinner panel for step 19.

14. Rubber door hinges are next. Cut two strips of rubber 3 inches wide and 1 inch longer than the door opening is wide. Inner-tube rubber is just right for the job; try to get a damaged truck or tractor tube, as it will lie flatter than a smaller car inner tube can.

15. Using a staple gun (or ½-inch roofing nails), fasten one strip to the front of the door piece

door. Mark the point that is exactly halfway between the peak and the center of the bottom edge of the wall. That will be the center of the opening.

10. Using a compass (if you don't have one, a thumbtack, string, and pencil will work), draw a circle of the right size for the opening (see page 51)

around the center point you have marked. The door needs to have an inch of clearance all around its frame, which means that a 12-inch diameter opening will require a 10-inch diameter door. Draw the smaller circle, using the same center and a radius 1 inch less than the first circle's.

11. Using a square (or a

and the other to the back of the door piece along the top, straight edge. Only a band 1 inch wide is fastened to the door, and the rubber will extend 1½ inches at the sides of the door. (When the door is in place, the strips will extend ½ inch beyond the opening at either end and 1 inch beyond the 1-inch gap at the top edge.)

16. Cut a U-shaped piece of rubber 2½ inches wide for the curved edge of the door piece. Fit it to the door so that a 1-inch band is fastened to the door and a 1½-inch band extends beyond the curve all around the edge. The top end of the U should butt up against the bottom edge of the straight rubber hinge that is attached to the straight edge of the door piece. Nail or staple it in place.

17. Lay the front wall on a flat surface and carefully center the door piece in the opening, maintaining a 1-inch clearance all around. The rubber strips should overhang the wall by 1 inch at the top and a ½ inch along the curved edge of the door.

18. At the top edge of the door piece, where you have fixed the rubber strips on both sides for the hinge, fit the strip on the back top of the door piece so that it overlaps by an inch the wall on what is to be the inside of the doghouse.

19. Nail the ¼-inch plywood door-shaped panel (from step 13) to the outside of the door, sandwiching the rubber strips between it and the door panel. The function of this outer panel is to hold the rubber strips firmly in place.

20. Nail the two lath strips over the portions of the straight rubber hinge that are fixed to the wall over the opening, inside and out, creating a sandwich similar to the one in the door. The door is hung. Trim the rubber later, as outlined in step 28.

21. Nail the front wall in place. Be sure to nail the bottom edges of all the walls to the floor all around, and the front and the back walls to the ends of the side walls and into the studs at all four corners.

22. Cut out the roof pieces; allow 1½ inches of overhang on each edge, and more if you want to give more protection. For the pitch of the roof on the house illustrated, a cut 12° from the vertical would be just right for the joint at the peak. If it isn't convenient to make the bevel when cutting, do it afterwards with a plane.

23. Attach two hinges as shown in Fig. 14.

24. Cut out a piece of rubber wide enough to extend down each side of the peak a distance of 2 inches and attach it along the ridge with staples or roofing nails.

25. Place the roof on the walls and center it carefully. Mark with a pencil all around where the outside walls meet the roof underside. Remove the roof and using box nails, nail the two long strips of lath to its underside so that the inner edges of the lath will just butt up against the outer sides at the top of the walls, thus both locating the roof on the walls and providing a seal against drafts. Before tacking the lath in place you might bevel the edges that will touch the outside tops of the walls so that there will be a snug fit, see Fig. 16.

26. You are finished with the building. Paint inside and out, including the rubber strips, with a good exterior enamel. The paint will protect the rubber from the sun's destructive rays. Be sure to let the interior air thoroughly for several days before you ask the dog to take up residence.

27. Replace the roof on the walls and screw in the hooks and eyes as shown in Fig. 16. Notice that the roof hooks each have two eyes so that either half of the roof can be propped open for ventilation. The whole roof, of course, may be removed for cleaning. This is also the time to screw in the hook and eye to hold the door open in nice weather. Use a 6-inch hook, screwed into the front wall about five inches over the door. Locate the eye on the door so that when the door is hooked up it is fully opened, but the hinge is not stretched.

28. Trimming the rubber weatherstrip on the door must be done with great care. On the one hand, the door must be able to swing in and out as pushed by the dog, and it should come to rest each time in a perfectly vertical position with the weatherstrip lying tight against the front wall. Some overlap, is, of course, necessary to provide a weather seal, but not much. Take your time and trim off narrow strips successively until the desired compromise is reached.

INSULATION

If you live in an area which has severe winters, or if you think your dog's ability to be comfortable outside in cold weather might be marginal, then you may well want to provide further interior insulation. A very good material for this is ordinary, inexpensive, indoor-outdoor carpeting, the kind with foam-rubber backing. You can easily glue panels of this carpeting, cut to fit, to the floor and to each of the two ceiling panels. Brown furniture glue applied generously works well for this.

On the walls, it will probably be adequate to glue it to the interior wall surfaces between the studs, although you may, for particularly good protection, choose to make up interior walls out of hardboard or ¼-inch plywood, and apply the carpeting to these. Put the interior walls in with clips or screws accessible through the carpeting. That way you can take them out at cleaning time and hose or scrub them off. Do not, by the way, fill the spaces between exterior and interior walls with insulation, unless you are very careful to make the spaces absolutely inaccessible to water. If water should get in and wet the insulation, it will stay there forever, possibly bring about mildew or mold growths, and will eventually rot out the wood.

GARDEN
SHEDS

The Need for a Shed

If you bark your shins on the lawnmower when you get out of your car in the garage, or if the wheelbarrow parked in the laundry room is beginning to generate adverse comment, the chances are you need a storage shed. You probably already know this, but you have your backyard just about where you want it, landscaped and planted, and the last thing you want to do is clutter it up with an ugly shed. But a storage shed, properly thought out, can be not only a handy thing but an aesthetic treat as well, and if you do your planning cleverly, it can be made to fit into just about any landscaping scheme.

APPROACHES TO DESIGN

There are two approaches you can take to the design of a garden shed. You can disguise it, or you can flaunt it, make it a center of attraction in your yard.

You can flaunt your garden shed by accentuating its size, its color, its form, or its situation. A large garden shed is hard to overlook. A large, brightly-painted shed is even harder to miss, especially if it is painted with vivid primary colors in an imaginative design. An unusual shape can also be eye-catching. One of the charms of plywood as a construction material is that it can be sawed into any kind of shape, so there's no reason why your garden shed has to have only right angles and conventional horizontal and

Courtesy of the American Plywood Association

A simple-to-build, no-frills garden shed. The ramp is handy for getting the mower in and out, and the racks built into the door are a good idea.

This building, made almost entirely of used materials, has in it elements of gazebo, garden shed, and greenhouse. Note how the upstairs front window appears to be a double-hung sash window—fitting to the structure's style—but is actually a pair of awning windows, hinged at the top. The problem of covering a curved roof is neatly solved with sheets of flexible galvanized metal roofing.

vertical planes. If you like the intricate scroll work, wood-turning, and finials of the Victorian period, you can accentuate whatever form you choose with that kind of decoration. See the suggestions for doing scrollwork on page 191. A prominent location will also serve to call attention to your shed. If you make it a focal point of your landscaping scheme, eyes will be drawn to it.

You can disguise your garden shed by blending it into the landscaping. If you locate the shed near a flowerbed, a trellis for climbing roses or morning glories is ideal, placed so it screens the shed from the view you want to protect. If the spot is in deep shade, or if the surroundings are more leafy than flowery, think about a trellis with English ivy. A grape arbor can screen a shed, and if the shed is north of the arbor, the sun reflected from its walls will do beneficial things for the grapes. If your garden is a formal one, a fruit tree espaliered against the wall of the shed could give it the look you want. Most books on garden architecture have instructions for trellises and espaliers.

If you have a largish back garden with a fair number of trees, you can make the shed unobtrusive by blending the construction materials with the surrounding flora. You might find that a simple clapboard shed, stained dark, will suit you just fine. If the weathered look appeals to you, there are weathering stains which gray with age yet seal the pores of the wood. Completely unprotected wood, provided it is solid wood and not plywood, will weather naturally and last almost indefinitely as long as the air is allowed to circulate freely on both sides; but if you let a plant, such as an espaliered tree or English ivy, grow against the shed, the wood will rot in time.

PLANNING FOR YOUR NEEDS

There is no reason why your garden shed has to have only a single function. The most obvious combination, of course, is a garden shed and greenhouse. If you build in suitable partitions, a manger, and ventilation, a conventional storage shed becomes a goat or pony stable. Add a window, a workbench, and storage compartments for seeds, bulbs, and pots, and you have a potting shed. With no interior modifications, all or part of the shed can be used to store

This simple wooden shed blends nicely into a corner of the backyard. After a few months, unpainted pine like this turns a silvery-gray which many people like, and which needs no maintenance.

firewood. Trays and bins along the walls are all you need for a root "cellar." (Just remember that the best preventive for food spoilage is free circulation of air, so don't put any trays on the floor itself.) You can also arrange a

workshop for your hobby or even a changing room, if you put the shed near your pool.

But to most people, a storage shed in the garden is a place to keep garden equipment. If this is what you have in mind, give

Screws from left to right are: hex-head plated lag screw (or lag bolt), square-head plain lag screw, pan-head sheet-metal (self-tapping) screw, slotted flat-head wood screw, Phillips-head combination (wood or sheet-metal) screw, smaller flat-head plated wood screw, round-head wood screw.

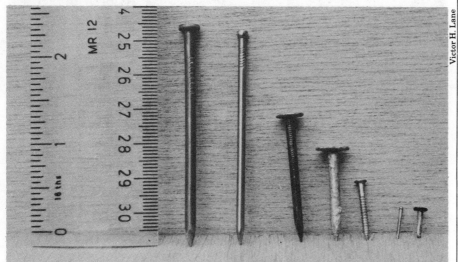

Nails from left to right are: common nail, finishing nail, box nail, galvanized roofing nail, serrated (or threaded) nail, brad, and tack.

some long and serious thought to how you are going to organize the stuff that is to be stored in the shed, and just what purpose beyond storage the shed is going to serve.

First, look at the size and shape of the items you want to store. For example, ladders can best be accommodated over head, slung from the ceiling with a pair of pulleys and ropes. Hoes, spades, rakes, and all other tools with long handles are best hung on the wall; they can be close together if they are hung at staggered heights or if every other one is hung upside down. Heavy things do best on the floor, so leave open floor space for the machines—mower, tiller, snow blower, what have you. If you decide to build cabinets for the machines, be sure to plan for adequate ventilation, so gasoline fumes don't collect. The fuel itself should be stored in a well-aired place, preferably one with a ventilated floor.

The best place to start is with an inventory of all the tools you plan to store in the shed. Measure the large items. You don't want to make the doorway too narrow for the wheelbarrow or mower. Plan the floor space and the traffic patterns so that mower handles and the like won't impede your movements. A good way to do this is to cut out outlines of the floor items from graph paper, using a scale of a half inch to the foot. Don't worry about all the little ins and outs of each machines or object; go for maximum measurements, but don't forget handles! Put these cutouts together on a fresh piece of graph paper, and you will have an idea of the kind of space you will need. In your planning, remember to allow for the thickness of the frame and wall.

A good thing to keep in mind is that there are bound to be items as time goes by that you will want to have room for in the shed. If you have a toddler now, a bicycle isn't far off. If you have a walking mower now, better ask yourself if there is any chance you may want a larger, riding mower in the future. Don't try to plan for absolutely everything, but neither should you plan only for those items on hand at the time of building. Also, leave some additional room for all those little things that seem to accumulate almost by themselves—a half-used bag of cement, pieces of left-over lumber, pots, paint cans, etc.

You may be limited in the size of shed you can build by zoning restrictions. Many municipalities strictly control the permissible size of outbuildings—something you should check into. Remember that to a surprising degree you can compensate for external size limitations with clever organization at the planning stage.

Owner-built wooden storage sheds usually use one of two basic construction methods: pole-frame or house-frame. Both are satisfactory, although in extremely cold regions, where there is substantial frost-heaving of the ground, or on sites that are not level (and not easily made level), house-frame construction is preferable. As a pole-frame shed requires no foundation of any kind, frost heaves could upset the entire structure. Since pole-frame construction is simpler, we'll deal with that first.

Steel storage buildings can be bought in a huge variety of sizes, with or without built-in floors and shelves. Most have permanent exterior finishes and sliding doors. Bought unassembled, they are easy to put together with an adjustable wrench and a screwdriver.

Building a Pole-Frame Shed

Here are directions for an 8x8-foot pole-frame shed. If you are going to sheath the building with plywood or particle board, which come in 4-foot-wide panels, it's handiest to plan the walls in multiples of 4-feet. The poles themselves are generally available from lumberyards; they are about 4 inches in diameter, tapering slightly, and have been treated to resist insects and rot.

This shed has an eave height of 7½ feet from the ground. As the floor height will be about 6 inches, you will have an interior height at the side walls of about 7 feet. For a shed, that is generous headroom, and it allows you to use 7½-foot precut studs, which are usually relatively cheap to buy. Eight-foot eaves are also, however, an easy option. You would need 8-foot studs, but you can use standard 8-foot plywood sheathing sheets without cutting, which you may find a significant convenience. Consider, though, that 8-foot eaves will give your shed a cubic look which you may not find aesthetically pleasing.

A structure as large as a shed is built in clearly definable stages, so instructions are given in stages—pole frame first, the floor, and so on. Each stage has its own materials and construction steps. Since there are choices along the way—whether to have a floor of gravel, concrete, or wood, for instance—you'll have to read through all the instructions and compile your total materials list before you start, to avoid repeated trips to the lumberyard.

POLE FRAME
Materials Needed
eight stakes and about 50 feet of strong string (to lay out the perimeter)
eight flat rocks about 5x5 inches (to support the bottoms of the poles)
eight building poles, at least 9 feet long, about 4 inches in diameter
sixteen 7½-foot 2x4s (to nail to the sides of the poles)
one 3½-foot 2x4 (end rafter stud)
2 cubic feet of 1½-inch screened rock (to fill in around the poles)
ten 6-foot 2x4s (for rafters)
one-half sheet (4x4 feet) of ¼-inch plywood (for rafter gussets)
eleven 8-foot 2x4s (for plates and stringers)
3½-inch (16d) common nails
twenty 4½-inch common nails
1¾-inch twisted or serrated siding nails
Note: A posthole digger is necessary for this project. Check the Yellow Pages for an equipment-rental company. Also, you will need a framing square.

WALL FRAMING
Construction Steps
1. Check or level the ground by laying the thin edge of a straight board on the ground and then laying a carpenter's spirit level on top of it. If necessary, shave the earth with a spade and fill as needed.

2. Lay out the perimeter of the shed with string and stakes, as shown in Fig. 17. Be careful that the area enclosed by the string is just 8 feet; certainly no more, if possible no less. Also be sure that the square described by the strings *is* a square. If it is perfectly square, the two diagonal measurements should be identical. Doublecheck these two points. A bit of care here can save you lots of trouble later.

3. Using a post-hole digger, dig holes just inside the perimeter string. In fact, the outside edges of the holes will have to extend beyond the perimeter strings just a little, so that when the pole is centered in the hole, with enough room around it for rock filler, it will stand perfectly vertically and just touch (but not press out!) the perimeter strings.

4. At the bottom of each hole set a flat rock large enough to support the whole end of the pole.

5. Set the pole into the hole, resting on the rock. Using the leveling board, mark ground level on the pole.

6. Remove the pole, measure from the ground-level mark, and mark the eave height. Cut the pole there. (See the discussion of eave height on page 66 .

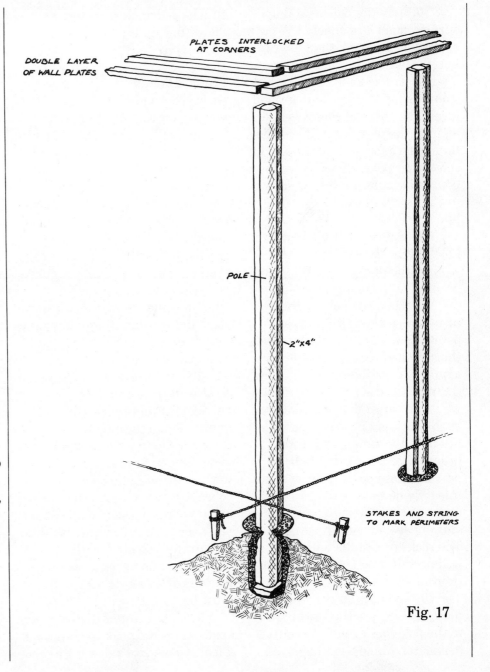

DOUBLE LAYER OF WALL PLATES

PLATES INTERLOCKED AT CORNERS

POLE

2"x4"

STAKES AND STRING TO MARK PERIMETERS

Fig. 17

7. Lay the pole on a flat surface, such as a sheet of ⅝-inch or thicker plywood, then lay two lengths of utility-grade 2x4 against it. The 2x4s should extend from the top of the pole to ground level. The edges of the 2x4s against the plywood will provide a flat surface for nailing the siding; the corner poles should have the boards at right angles, as shown in Fig. 18. Nail the boards in place.

8. Set the pole back into the hole so that the side that was against the plywood is facing and parallel to the string.

9. Hold the pole vertical (use a spirit level to check) and pour into the hole around the pole the screened rock, available from a quarry or builders' supply yard. (Do not suppose poured concrete would be better than the flat rock at the bottom and screened-rock filler. As the siding and rafter braces, discussed later, will prevent these poles from swaying, it is only necessary to keep them from shifting laterally and sinking deeper into the ground. The rocks accomplish all this, and will permit much better drainage than would concrete.)

10. Now lay on the wall plates, which are the 2x4s nailed to the pole tops around the entire perimeter. Make a double layer, interlocked at the corners, as shown in Fig. 18.

RAFTERS

No matter what grade of lumber you use for the building, don't skimp on the rafters. Get the best grade of construction lumber your lumber yard has to offer, and as you buy, check each board for warp in either plane, and for flatness. Be ruthless in rejecting boards, look for twists and cracks. If the finished rafters are even a little warped or irregular, you will have serious difficulties when it comes to laying down the roof, and you can imagine what a longitudinal crack does for the strength of a rafter!

To cut rafters accurately, use a rafter framing square, also called a rafter square. Good-quality squares come with directions for cutting rafters. It is not a particularly difficult job, but one that can be hard to conceptualize without the square and a board to practice on. What you need to do, of course, is make cuts in the wood that will be horizontal and vertical when the board is at an angle, as a rafter must be.

If the *horizontal* distance, or span, from roof peak to wall top is 4 feet or less, 2x4 rafters will be strong enough, even if heavy snow can be expected in your area. If little or no snow falls in your region, 2x4 rafters can be used for spans up to 7 feet. For longer spans (but not more than twice as long), use 2x6s.

Construction Steps

1. Cut the rafters—both the angle for the peak and the notch near the other end—following the directions for rafter cutting that come with the rafter or framing square.

2. Plan to space the rafters 2 feet apart, center to center. (If you need extra strength, set them on 16-inch centers.)

3. Put the rafter pairs together on the ground first. To join the abutting rafters at the top, make gussets of ¼-inch plywood, as shown in Fig. 18. You can also use metal gussets available from a builders' supply store. If you make your own gussets, a quick way to do so is to cut your ¼-inch plywood into 2-foot squares, and cut each square into four gussets by slicing along the two diagonals.

4. Nail a gusset securely to the peak end of *one* rafter board. Use galvanized twisted or serrated siding nails, 1¾ inches long. Use at least 9 nails.

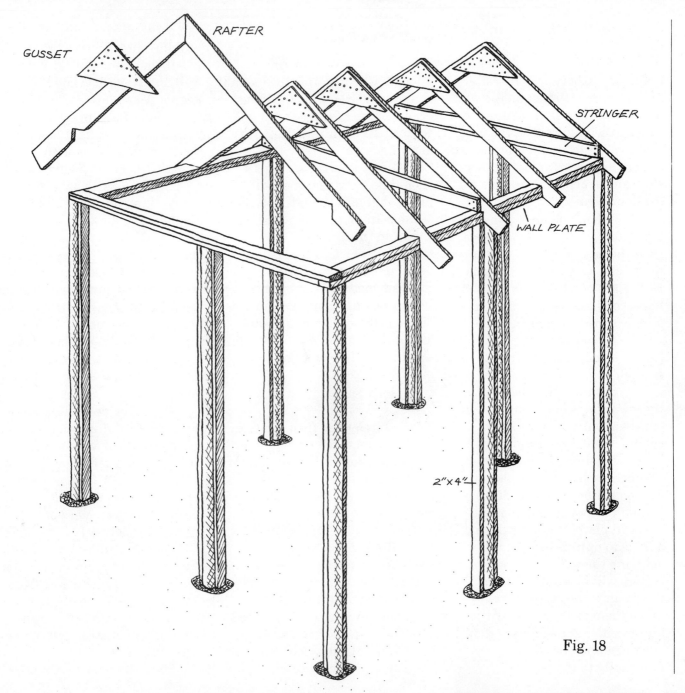

GUSSET

RAFTER

STRINGER

WALL PLATE

2"x4"

Fig. 18

5. Set the other rafter board in position but don't nail it in place until you carefully measure the distance between the rafter notches that will rest on the wall plates. A very small mistake in sawing the angle cut at the peak of the rafter or in butting the two rafters together can result in a substantial error in the span of the double rafter.

6. When you have joined the two rafters with the first gusset, turn the assembly over and once again measure the span between wall-plate notches before nailing the second gusset. The reward for a little time taken at this stage will be a good, straight, flat roof. Do the same with each rafter assembly.

7. Set the rafter pairs onto the wall plates one by one as you get them built. That way you will catch any mistake before you have compounded it. Use two 4½-inch nails to fasten each of the ten rafters in place, driving through the rafter notches into the wall plates. Stagger the nails a little to minimize the danger of splitting the rafter. As the rafters will be vulnerable to winds until the roofing is on, brace them each temporarily, either with boards joining them one to another, or running diagonally to the plate.

STRINGERS

You are now ready for the stringers—2x4s laid on edge that span the width of the building. Each stringer forms a triangle with a pair of rafters. Refer to the drawing.

The ends of the stringers should lie firmly on the wall plates, but nail them to the side of the rafters, not the plate, as shown in the drawing. One stringer every four feet is the usual prescription. For an 8x8-foot shed you will need 3 stringers. (You can increase the distance between stringers to 6 feet, if the length of your building is not a multiple of four.)

FLOOR

Next consider the floor. You can, of course, omit flooring completely and get along with the bare earth, but if you want to keep stored items dry, that isn't a very good idea. Even on a level site, some water will seep under the walls of a pole shed, and there is also rising damp from the earth to contend with. Three good floor options are crushed rock, poured concrete, and wood.

The first two are both easier and cheaper than wood, but only if a delivery truck with materials can drive right up to your building site. Crushed rock also has the advantage that it drains well and stays dry, and if you keep it raked level, it looks very tidy. The disadvantages are that you do have to rake it now and then, and machines with small wheels, like most lawnmowers, won't roll properly on it. If you are working on one of your machines on a rock floor, and drop a little part, it can very easily disappear forever among the stones. One solution is to lay ¼-inch hardboard down loose on the rock. It makes a strong, dry floor, and you won't lose little objects so easily—although some people will find the looseness bothersome, if more for aesthetic than utilitarian reasons.

If you decide on a concrete floor, the first thing you must do is figure out the volume of concrete you need in cubic yards. A floor 5 inches deep and 8 by 8 feet in area, for example, will require one cubic yard of concrete. Ready-mix companies will usually deliver any quantity of a cubic yard or greater, and if you can get your concrete from them, it is the easiest way to do the job. For mixing your own concrete, see instructions on page 98.

For either crushed rock or

concrete, you must do the following preparation.

SKIRTING FOR CRUSHED-ROCK OR CONCRETE FLOOR

Materials Needed

four 2x6s for skirting boards
(more may be needed; see step
2)
18-inch stakes made of 1x2
furring strips
3½-inch (16d) nails

Construction Steps

1. If you want a crushed-rock floor, nail 2x6 skirting boards (sometimes called splatter boards) around the *inside* perimeter of the pole framework, at ground level. If you want a concrete floor, nail the boards inside as in Fig. 19, page 72 or if you prefer, around the *outside* perimeter of the poles, at ground level. Be sure the tops of the boards are level, if that's how you want the concrete floor to be. If you want the floor slightly sloped for drainage, then slope the boards accordingly. About 1 inch in 8 feet is adequate.

2. Because the poles are 4 feet apart and a 1x6 is fairly flexible, the skirting board needs additional bracing. If your site is earth or sod, without a lot of rocks, make up twenty-four 18-inch stakes, pointed at one end, out of 1x2 furring strips; plant one at least every 16 inches along the outside edge of the skirting board. If the ground is very hard or fairly rocky, you may have better luck with cut and sharpened sections of steel fencepost. If the ground is very rocky or very loose, you will have to go to a heavier skirting board—either a side-by-side double line of 1x6s or a single line of 2x6s.

3. Fill the skirting with rock or concrete.

CRUSHED-ROCK FLOOR

One cubic yard of 1-inch to 1½-inch screened rock will generously fill the skirting if the floor is set within the poles of an 8x8-foot shed. Rake it smooth with a coarse garden rake.

CONCRETE FLOOR

Materials Needed

ten 8-foot reinforcing rods
(optional)
baling wire (optional)
1 cubic yard of concrete

Construction Steps

1. If you live in an area where deep and long frosts occur in winter, reinforce the concrete with steel reinforcing rods. You can buy the rods in 8-foot lengths from a builders' supply yard.

2. Using a solidly mounted vise, bend a 3-inch right-angle hook at each end of each 8-foot rod; as shown in Fig. 19.

3. Make a lattice of the rods by laying them in rows, 2 feet apart, both north-south and east-west.

4. Bind each junction together with a twist of baling wire, tightened with a pair of pliers. The hooks should point downwards so as to elevate the whole lattice by 3 inches. If it sags anywhere, prop it up with stone. The exact height of the lattice isn't important, so long as concrete can run both under and over it.

5. When the concrete is poured, spread it with a sturdy rigid-tined garden rake. Be ready with it when the mixer truck arrives, so you can spread as he pours. When spreading, turn the rake so the tines point upward. This way you won't hook them on the reinforcing rods. Be careful in general not to disturb the network of rods.

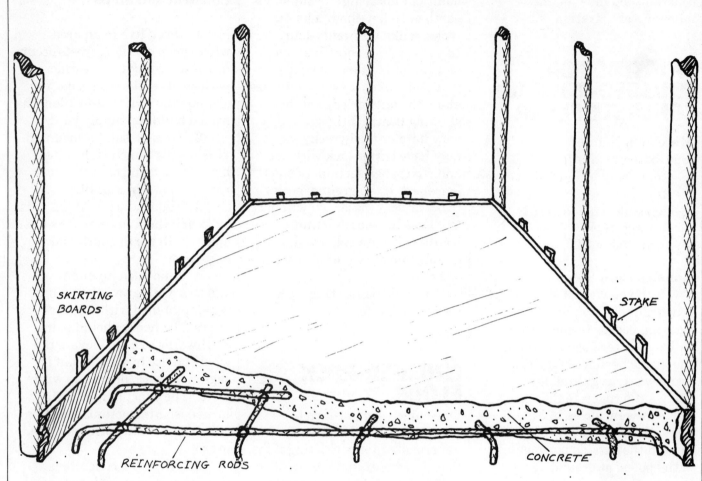

SKIRTING BOARDS

STAKE

REINFORCING RODS

CONCRETE

Fig. 19

6. When the concrete is pretty well spread out, turn the rake around and with the handle end poke the wet concrete all over, to work out air bubbles. Again, don't be too vigorous, or you may make a hash of your reinforcements.

7. Smooth the surface with the edge of a flat board, if possible one the right length to rest on the skirting boards on both sides. This process, together with the poking to remove air bubbles, is called screeding; the tool with which you do it is called a screed—even if it's just an upturned garden rake.

8. Let the floor dry at least 24 hours before walking on it. You'll have to remove skirting

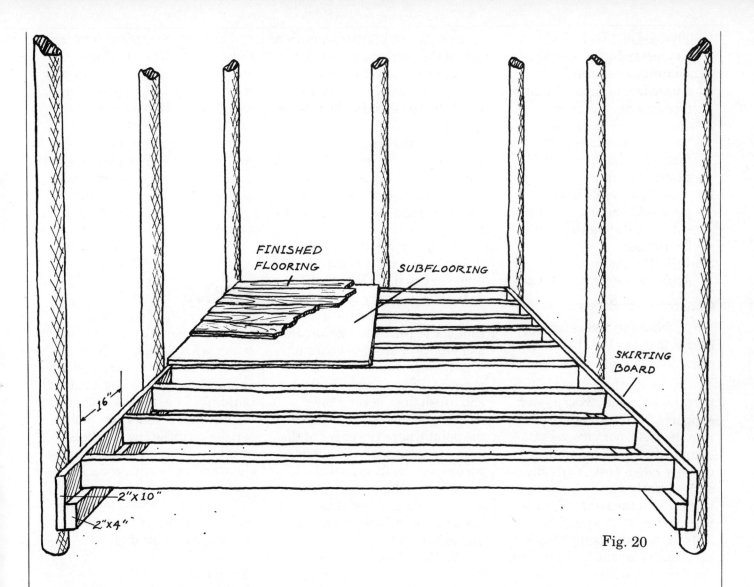

FINISHED FLOORING

SUBFLOORING

SKIRTING BOARD

16"

2"x10"

2"x4"

Fig. 20

boards before putting on the siding.

WOOD FLOOR

To build a wood floor, here's what you must do.

Materials Needed

four 2x8s (you may want to use 2x10s), 8 feet long (skirting boards)

five 2x6s, 8 feet long (joists)

two 2x4s, 8 feet long (joist supports)

64 square feet of subfloor, either ⅝-inch subfloor-grade

plywood or 1-inch-thick tongue-and-groove or shiplap boarding (see step 7)

64 square feet of finish flooring (see step 8)

3½-inch nails or thirty-six ¼x4-inch carriage bolts to fasten joist supports to skirting boards

1¾-inch siding nails to nail down plywood floor or 2-inch (6d) siding nails to nail down a board floor

3½-inch (16d) nails to nail skirting boards to joist ends

Construction Steps

1. Nail the 2x8 skirting boards to the poles, their bottom edges 2 inches above the ground.

2. Along the bottom of two parallel 2x8s, nail 2x4s as joist supports. Nail through the 2x8s into the 2x4s, not vice-versa, except where you can nail directly into a post. Nails should slope slightly upward, so weight will tighten the joint. (If you feel you need extra strength, bolt the 2x4s to the 2x8s using ¼-inch carriage bolts, put through from the 2x8 side, with flat washers under the nuts, and set a foot apart.)

3. Cut floor joists of good-quality 2x6 lumber to fit inside the 2x8s, ends resting on the 2x4s. The top of the joists will be an inch or a bit more above the tops of the skirting boards. (If you want the top of the joists flush with the skirting boards, use 2x10s for skirting boards, as Fig. 20.)

4. Set the joists 16 inches apart, center to center. (This arrangement, using lumber of the dimensions given, will be adequate for a floor span of up to 12 feet.)

5. Nail the joist in place through the 2x8; it is not necessary to nail them to the 2x4s on which they rest.

6. On the joists lay either a subfloor of 1-inch-thick tongue and groove or shiplap boarding, or of ⅝-inch subfloor-grade plywood. Extend both subfloor and finish floor beyond the skirting boards on all sides, to the point where the edges of the floor are flush with the outside surfaces of the 2x4s nailed vertically to the poles. In this way the siding, when it is applied, will exactly meet the edges of the floor.

7. Nail each course of flooring to the joists, using specified nails, set every 8 inches on plywood, or two to the width of one board. Use a carpenter's chalkline to help find the joists when they are covered by flooring. Locate the joist at either end and run the string between those points. Then hold the string tight and give it a snap. It will lay a chalk line down the length of the hidden joist.

8. On top lay the floor. It may be boarding, like the subfloor, but laid at 45° or 90° to the lower course of boards, or it may be one-side-good ⅝-inch tongue-and-groove plywood. Nail as above.

SIDING

There are lots of ways to side a pole shed, but best, I think, is either with ½-inch exterior plywood or a good ⅜-inch particle or chipboard. I have used both in buildings I put up in Canada and Nebraska ten years ago, and I can see no difference in their durability yet. The chipboard is cheaper, but be sure you get exterior grade, from a reputable lumber yard. I have seen some distinctly inferior cut-rate particle board around.

Materials Needed (8x8-foot shed)

nine sheets (4x8 feet) siding, either ¼-inch exterior plywood or ⅜-inch good-quality chipboard

twelve 8-foot 2x4s for nailers

1½-inch serrated or threaded
    siding nails
3½-inch common nails (for the
    nailers)

Construction Steps

1. Use the sheets of siding
vertically. Cut them so they tuck
firmly up against the lower edges
of the rafters and come to an inch
from the ground. Nail the sheets
to the 2x4s you earlier nailed
against the pole sides, not to the
poles themselves, which are less
likely to present a flat, even
surface. Leave one piece of siding
off so you can get in and out. For
reasons that will become clear in
the next construction stage,
make it a piece that is *not* to have
a window or a door in it.

2. When the vertical sheets
of siding are all (but one) on,
there will remain only the gable
peaks to do. Here is an easy way
to do them one at a time: When it
is still uncut, tack-nail (that is,
don't drive the nails all the way
home) the last sheet of siding
horizontally across the gable end,
its lower edge resting on the
upper edge of the siding below it.
When it is up it will look like the
false front of a wild-west store
building. What you want to do is
leave the siding but remove the
false front, so using the top of the
end rafter pair as a guide, mark a

line on the siding. Take the
siding down to cut it (too
awkward to saw it in place, and
too much danger of cutting into
the rafter), then put the piece
back up and nail it firmly. Be
sure the end rafter is solidly
braced before you hammer
against it!

3. You are now left with
piece of siding shaped like a
flattened letter "M." Set it aside
for the moment. Take the 3½-foot
2x4 gable stud up to the
still-open gable end, set its base
on the plate directly below the
peak, and hold its upper end
against the peak. Using the
lower side of the rafter pair as a
guide, mark the stud so that
when cut it will just fit vertically
between the peak of the end
rafters and the plate. Take it
down, cut it, and then toenail it
in place, top and bottom.

4. Cut the "M"-shaped piece
of siding in half, vertically
through the bottom of the dip of
the "M." Reverse the pieces right
to left, and you have the
wherewithal to side the
remaining open gable end.
Tack-nail the two pieces in place,
mark the cut lines (at the
bottoms of the pieces this time),
take them down and cut them,
and nail them firmly in place.
Apart from the piece you have

intentionally left off, the siding
is done.

WINDOWS AND DOORS

Most people will tell you to frame
in the window and door openings
before you put up the siding, but
with a pole frame and sheet
siding, I find it easier to do it the
other way around.

Careful planning is the first
essential. You have already
planned the interior space
allocations. If you did not include
door placement in those plans,
consider it now, in light of the
space usage you have projected.
Be sure the location of the door
doesn't force you to walk over
floor space you need for storage of
a particular piece of equipment.
Figure the door width you will
need for, say, your riding mower
or the wheelbarrow. And decide
whether you want the door to
swing in or out, hinged on the
right or the left. A consideration:
A door hinged to swing out won't
bump into stuff inside the shed
when you open it, but it is harder
to lock securely than an
inward-swinging door.

Windows in a garden shed
usually are there only to admit
light, and occasionally air. You
seldom need to see out, and that

fact can have an important advantage for you. If your aesthetic considerations permit it, you can decide on one or more high, horizontal windows, lying just under the eaves, which will leave your wall space available for hanging tools and things. You may also not need the added complication or cost of opening windows, although you should remember that in the summer a slightly open window can do a lot of forestall mustiness and mildew.

SOURCES AND SIZES

After you have determined where you're going to put your door and window units, but before you do any cutting or framing, you must know the exact requirements for each opening. Only if you plan to make your own units (see the instructions on pages 80 and 81) can you count on making the unit fit the opening.

The cheapest window, of course, will be one you get for nothing, which means a used window. The easiest ones to find, generally, are old wooden storm windows, now replaced on many houses by aluminum combination windows. Check with your neighbors.

You can often buy used windows, too. If there is a used-lumber dealer or house wrecker in your area, check there.

Windows of virtually any size and type may be had on special order from a lumber dealer, but the cost of special-order windows may be triple the price of a similar stock size. So find what's available locally.

There are basically two kinds of windows that open—those with sliding sashes and those with hinged sashes. The most common type of sliding-sash window is the double-hung unit, whose upper and lower sashes slide independently; in a horizontal sliding window, commonly only one sash slides sideways, the other half being fixed in place. The casement window is the most common hinged type, with the hinges at the side; other styles are awning (hinged at the top), and hopper (hinged at the bottom). Sliding-sash windows are easier to protect with screens and storm windows than the other types.

Double-hung windows are not generally catalogued in sizes smaller than about 2x3½ feet, overall. Casement, awning, or horizontal sliding units, however,

may be bought out of regular stock as small as 20 inches by 32 inches.

The largest single-unit double-hung windows available in stock are about 4 feet wide by 6 feet high. Other window types are commonly available up to about 3x4 feet or 2x6 feet. Intermediate sizes are usually found in width increments of 4 inches and length increments of 8 inches.

By the way, don't overlook the so-called basement or attic windows—some of them are very nice, particularly the ones that slide open and have the screens already in place.

As for doors, used ones are not quite as easily found as used windows, but used-lumber merchants generally have them.

New doors, paneled and plain, with and without windows, can be ordered from stock in widths from about 2 feet to about 3½ feet, usually in 2-inch increments. Standard heights are 6 feet 8 inches for interior doors and 7 feet for entrances. These, by the way, are the sizes of the door and its opening, not the overall sizes of the whole unit including door jambs, etc. For those, and for the sizes needed for the rough opening in the wall, check with your dealer.

CUTTING AND FRAMING THE ROUGH OPENING

The opening in the frame and siding, called the rough opening, should be about ½ inch wider and taller than the overall dimensions of the window or door unit. Your supplier will give you precise recommendations, but the exact figure is not so important as long as you don't have to squeeze the unit in place or leave an exaggerated gap around the frame.

Measure and mark the openings you want to cut out. Measure twice, and make sure all corners are square.

The only trick to cutting out the opening is in getting the hole started (see box on page 106). If you're using a power saw there's no problem, as they are capable of starting their own holes. If you are using a hand, keyhole or compass saw, drill holes through the siding just inside each corner, large enough to admit the blade. You may want to switch to a larger-bladed saw once the cut is long enough to accept one, as with a larger blade it's usually easier to cut a straight line, and it's always faster.

Framing is shown in Fig. 21. It is made of 2x4s, and is similar

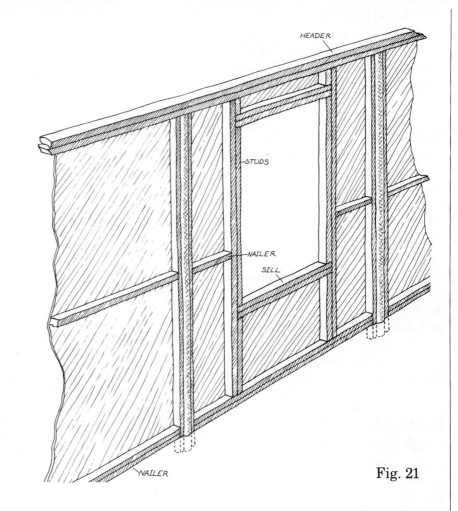

Fig. 21

to what is used for house-frame construction, except that the bottom element is the double nailer not a conventional single shoe.

Measure for and cut the nailers. Be sure to measure separately for each nailer, as there is bound to be some variation in the spaces to be spanned. Toenail the doubled bottom nailer between the studs nailed to the poles, about 4 inches from the ground or, if you have laid a floor all the way out to the siding, then just above the floor. The intermediate nailer should be placed halfway between the lower nailers and the wall plates.

Note that the framing is doubled all around the window opening, and that the inner pair of studs is interrupted to carry the horizontal window sill and header.

If you are not using a complete commercial window unit, you should leave room at the bottom of the opening for a sill. You can either buy one ready-formed at a lumberyard, or make one by tilting a 1x4 or 1x6 slightly up on the inside, supporting the raised edge with a couple of shims (see glossary).

INSTALLING READY-MADES

Now is the time to set whatever window unit you have chosen into the rough opening. Fix it in place by wedging shims between the unit and the rough framing. If yours are standard units, you will be able to nail through the frames into your rough framing.

If you are using old storm windows or something similar, fix the windows in place by edging them on all sides with

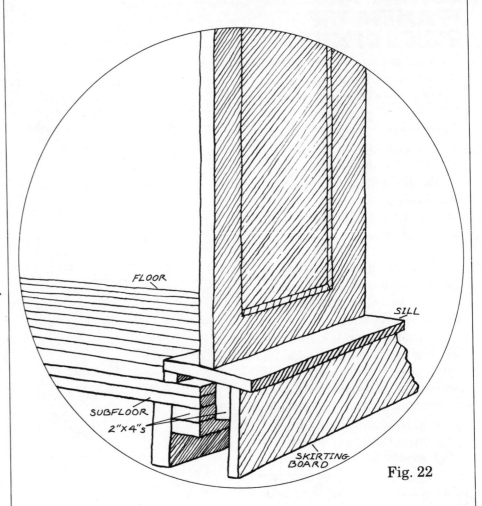

Fig. 22

molding (quarter-round or lath or 1x2 does nicely). Whether the window is fixed in place or designed to swing out, it's a good idea to have a sloping sill over the cut edge of the siding, to keep rain from seeping inside; in that case, molding is not needed on the bottom outside. And if the window is hinged, you will want molding only on the shut side; in that case it's called a stop. An awning window should open outward and a hopper inward, both for rain protection; a casement can open either way.

The same comments apply to the door. If you use a complete commercial unit, it will come with jamb and sill. The complete unit is put in place, adjusted, and shimmed, as with the window units. Then it is nailed through the door jamb into the rough framing. In a pole-frame building, you should give some extra support to the sill—unless you have a full-width concrete floor, in which case the sill can rest securely on that. Otherwise, the front of the sill should be supported with a skirt board which reaches the ground (if solid and firm) or, better, rests on a stone or concrete slab. See Fig. 22. If you make your own jamb and sill, you can arrange it just as you would if it were a window. See the drawing of a window and frame, Fig. 23.

When your windows and door are securely in place, you can nail up that last piece of siding, left off earlier for access.

Finally, check the whole building to be certain the siding is securely nailed to all the 2x4s: the pole studs, the nailers (middle and bottom), the wall plates at the top, and the window

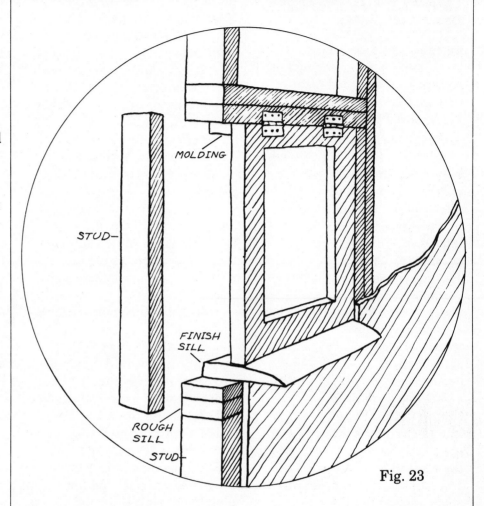

Fig. 23

and door frame members. A chalk line helps to locate the nails when the 2x4s can't be seen from outside. Space the nails about one hammer-length (head and handle) apart.

Below are instructions for making your own window and door. The installation tips apply also to other doors and windows that don't come as complete units.

SIMPLE CASEMENT OR AWNING WINDOWS

If you don't need anything fancy, you should have good results with simple windows made of 1x2, grooved to accept the glass. They can be hung either as casement or as awning windows.

Materials Needed

1x2 for window frames
1x4 to form frame in rough opening
1x6 for sill
1x2 for stop bead
2¼-inch finishing nails for outer frame
2¼-inch #6 flathead screws for window frames
carpenter's glue
pane of glass
hardware (two 2-inch hinges and a latch)

Construction Steps

1. Frame all four edges of the rough opening with 1x4s. Put the top and bottom boards in first, and cut the side boards short enough to fit against the top and bottom boards. The flat side of the 1x4 rests against the 4-inch side of the 2x4 framing, but it also covers the raw edge of the siding.

2. Nail on the sill, as shown in Fig. 23. If desired, the sill can be nailed directly onto the horizontal 2x4 forming the bottom of the rough opening, eliminating the 1x4 in that position.

3. Frame the opening with 1x2 to form a stop bead. The inside edge of the 1x2 should be flush with the inside edge of the 1x4 frame. Plane the bottom of the bottom stop bead so that it lies flat and horizontal on the sloped sill.

4. Measure the width and height of the full opening (formed by the 1x4s) carefully, subtract ⅛ inch, and cut pieces of 1x2 to those dimensions, two to each length.

5. Miter the ends of each piece to 45°.

6. With a table or radial-arm saw cut ¼-inch-deep grooves down the middle of the shorter narrow edge of each piece of 1x2. Be sure to set the saw guide carefully and tightly before you begin so that the grooves in all four pieces will match when the pieces are put together. If you plan to use single-strength glass, cut the grooves ⅛ inch wide; if double-strength, $5/32$ or $3/16$ inch wide.

7. Using corner clamps (fairly inexpensive at your hardware store, if you don't have any), clamp the frame together, making sure the grooves mate up perfectly. Use glue on both ends of the piece that is to carry the hinges, but on neither joint of the piece opposite it. Drill holes for the screws, drilling through both the longer pieces across the glued joints into the length of the shorter pieces. Screw tight the two glued joints only and set the frame aside, still clamped, until the glue sets.

8. Measure the inside opening of your frame, while it is still clamped. Buy a pane of glass exactly ⅜ inch larger in both dimensions than the inside opening of the frame. This extra amount will fit into the ¼-inch grooves, leaving a slight amount of play for expansion.

9. Unclamp the frame and remove the yet unattached side, but only after marking it so you can put it back just the same way.

10. Rub the edges of the glass well with a bar of soap, then slide it into the frame. Replace the fourth side of the frame and screw it into place, but *use no glue*. That is the piece you will remove if you ever need to replace the glass.

11. Putty the glass tightly on both sides so that neither rain (from the outside) or condensation (from the inside) can run down into the groove and cause rotting.

12. Hinge the window on the edge away from the edge that is detachable. That way you won't need to unhinge the window to replace the glass. Be careful, when you screw on the hinges and latch, not to drill or screw into the glass hidden in the groove.

MAKING YOUR OWN DOOR

All in all, there isn't a better shed door than a board-and-batten door. It's a lot nicer looking than a plywood slab, and it's simple as anything to build.

Materials Needed

shiplap or tongue-and groove boards of any width totalling at least ½-inch more than the width of the doorway, each board as long as the door is high
1x4 for the batten Z—two pieces the width of the door and one longer for the diagonal
1x2 for stop—two pieces the height of doorway, one the width
1¼-inch #6 screws *or* 2-inch common nails
1½-inch finishing nails
one thumb-catch latch
two 8-inch or 10-inch T-hinges

Construction Steps

1. Cut the boards ⅜ inch shorter than the height of the doorway opening. (You can first put 1x4s inside the doorway for a finished frame, as with the window, if you wish.)

2. Trim the tongue (or lap) from one side of one of the boards, and then lay the boards together on the floor with the trimmed edge as one side of the array. Measure a total width of ³⁄₁₆ inch less than the width of the doorway opening, and trim the excess from the other side of the array. Be especially careful that the boards are pressed tightly together before you measure.

3. Cut 1x4 for the top and bottom battens. They should be shorter than the full width of the door, so the battens don't get in the way of the stops when the door is closed. Depending on what boards are used as stops, the battens should be from 1 to 4 inches shorter.

4. Screw the battens to the boards, after drilling holes for the screws, making sure the boards are held tightly together. The battens go on the inside of the door, lying across the boards, 4 inches from the top and bottom of the door.

5. Carefully mark the third 1x4 for the diagonal batten. It should be on the hinge side at the bottom, running up toward the free edge at the top. The ends of this batten are cut at an angle so they will butt against the edges of the top and bottom battens. The function of this batten is to keep the door square; without it, the door would sag in time.

6. Check the door for size. You want to install it with ⅛-inch clearance between the door and the top jamb and latch-side jamb, ¼ inch at the bottom, and ¹⁄₁₆ inch on the hinge-side jamb. If necessary, trim some off.

7. Install the hinges and latch; use the manufacturer's instructions for the latch. When positioning the hinges, keep in mind that you will have to install a stop bead (next step).

8. Close the door in the latch. Cut one piece of 1x2 the width of the doorway. Nail it onto the top jamb with finishing nails, just touching the door. First tack the top in with just a couple of nails driven partway in, then check it for fit. If the door shuts and latches all right, do the same with the two side pieces of 1x2, which run from top to bottom of the jambs. If the door opens and closes easily, the stops are correctly placed. Finish nailing them in place.

ROOF

The framing is done. You have only to complete the roof and take care of some finishing touches.

The conventional roof covering is ⅝-inch plywood. Start at one end of the ridge and work down to the eaves. If you have to have a seam at right angles to the rafters (only necessary if the run of the roof on one side is more than 8 feet), nail 2x4 bridging between the rafters, under the seam, so the plywood won't be able to flex there, and the edges can be nailed down.

On top of the plywood nail the finish roofing. The usual choice is composition shingles, although you may want to consider something more exotic such as cedar shakes, red tiles, or slates.

Corrugated galvanized roofing sheets are very economical. For a garden shed, where excluding insects is a low priority, you can nail the corrugated metal sheets directly to the rafters, using ordinary roofing nails with soft washers that deform under the nailhead and seal against water. If a tighter or stronger roof is required, use an underroof of ½-inch plywood and nail the galvanized roofing on top of it.

During the season, this shed doubles as a roadside lemonade stand.

FINISHING TOUCHES

There are still a few finishing touches to take care of. You will be left with some small openings under the eaves, about the plates and below the roof sheeting, between each pair of rafters. If you want the shed to be tight, you must plug them, either from the inside or the outside. You can use either sections of 1x4 or pieces of scrap plywood, if you have any left over.

If you want to provide ventilation in the shed, leave one or more of the openings unplugged and covered instead with window screening, held in place with strips of lath and roofing nails. This is most easily done from the inside. The screening is important, by the way, even if you don't care about a few mosquitos or flies in your storage shed. Birds are the big problem. They will soon discover the appealing little openings, with inconvenient and unpleasant consequences for you.

If you have chosen a crushed-rock floor confined by skirting boards fastened to the insides of the poles, you will have a gap between the edge of the floor and the walls. Now that the siding is in place, you can fill those gaps with more of the same crushed rock.

You may also feel that the door and windows look a little bare on the outside, particularly if you have put in your own units. You can dress them up by nailing 1x4s flat against the siding, also covering most of the edge of the 1x4 finished frame.

House-Frame Shed

House-frame construction is the second major alternative. This method uses a framework made primarily of 2x4s, which must rest on a foundation. Many of the construction steps are the same as those used for the pole-frame shed, so you'll often have to refer to those directions.

FOUNDATION

Some sort of foundation is needed. For a storage shed, though, nothing very elaborate is necessary, particularly if your site is reasonably level. The structure can rest on several concrete blocks, or it can be mounted on posts sunk in concrete.

CONCRETE-BLOCK FOUNDATION

To make a concrete block foundation, which is sufficient if the ground is fairly firm, lay the blocks on their sides, holes out as shown in Fig. 24. Place blocks every 4 to 6 feet.

For best stability, the blocks should protrude beyond the house frame about 3 inches; take that into account when you measure for their exact location.

If the ground is not out of level more than the thickness of a concrete block, less a couple of inches, you can arrange a level foundation simply by digging shallow (but flat-bottomed) holes for the blocks to rest in. If the ground is a little soft, you can harden it nicely by tamping into it a course of 1½-inch screened rock. Use a long 2x6 or 2x8 and a carpenter's level to check whether the tops of the blocks are flush and even with one another.

If the ground is more uneven than can be compensated for by partly burying some of the blocks, you can use two blocks where necessary. If you stack only two high, no mortar is necessary for a simple storage shed 12x12 feet or less. But if you need to stack blocks more than two high, use mortar between them. If more than four would be needed, go to posts.

POST FOUNDATION

For a storage shed subject to normal use, a 4x4 cedar or redwood post will supply

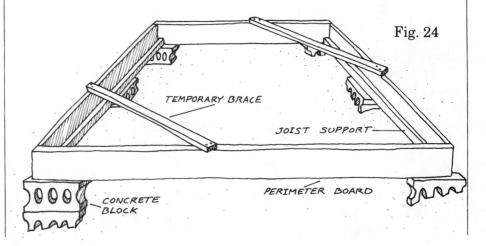

Fig. 24

TEMPORARY BRACE

JOIST SUPPORT

PERIMETER BOARD

CONCRETE BLOCK

adequate support, providing no more than 3 feet of elevation is required. If more lift is needed, you must go either to larger posts, or to braces for the narrower ones. The usual span is 4 feet between posts. Because conditions, needs, and materials vary so greatly, check with your lumberyard if long posts will be necessary to support part of your shed. It's a good idea, by the way, to start with a post that is several inches longer than you know you will need, including the part underground. You can saw the top later to just the right height.

The posts require more sophisticated work, both above ground and below ground, than do concrete blocks. The best way to set a post is in concrete.

Materials Needed:

posts as required, 4x4 or larger (see above)
concrete, about one cubic foot for each hole
1½-inch screened rock or coarse gravel, about ¼ cubic foot for each hole
1x2 or similar short boards for braces, two for each post
stakes and string to lay out shed plan. (A full-length perimeter board, to be used at this stage only for leveling.)

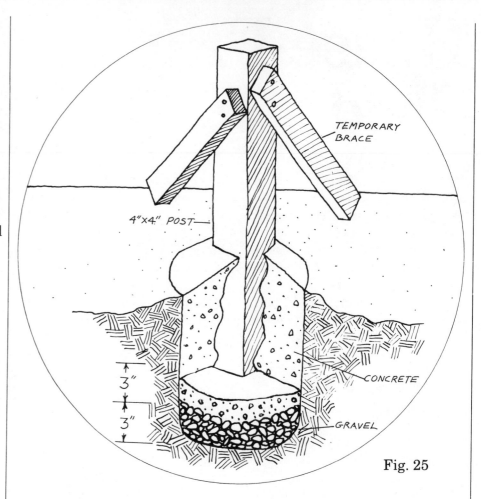

Fig. 25

Construction Steps

1. Lay out the perimeter lines with stake and strings, as for the pole-frame shed (page 67, Fig. 17).

2. With a posthole digger (available from equipment-rental companies), dig a hole 12 inches deep for each post.

3. Pour into each hole about 3 inches of rock or coarse gravel, then 3 inches of concrete, as shown in Fig. 25. Let the concrete set.

4. Recheck the stakes and strings, making sure the corners are square and the diagonals are exactly the same length.

5. Nail scrap boards onto

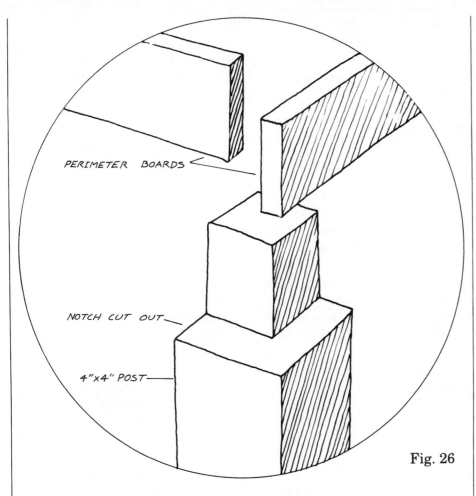

PERIMETER BOARDS

NOTCH CUT OUT

4"x4" POST

Fig. 26

The surest way to get them all level is with a builder's transit, which can be rented, with directions for use, from an equipment-rental company. If you are willing to settle for slightly less exactitude (and this *is* a garden shed), the best way to level is with a long, heavy, straight board. One of the shed perimeter boards would be just right—but sight along it carefully to be sure it's dead straight. Then saw the first post to the desired height. Set the edge of the perimeter board on the cut post, and lay the side of it against the next post in line. Place a carpenter's spirit level on the board, and when it is perfectly level, mark the next post. A good check, by the way, is *not* to cut the second post, but hold or tack-nail the board to it at the right level, and mark the third post, and so forth all around the perimeter. If you come back to the cut post at the right height, you'll know all the marks are at the right levels.

When you have the posts cut, notch the sides, as shown in Fig. 26. (Posts that are not on the corners should, of course, be notched only on the outside face.) Remember that levelness is, if anything, even more important at this stage, as it is on those

the posts at 90° angles to form temporary braces, as shown in the drawing. Put a post into each hole, hold it vertical (use a level) and just touching the corner of the strings. Drive stakes into the ground at the ends of the braces if needed to keep the posts vertical.

6. Pour concrete into the hole around the posts. Mound the concrete up a bit around the post so that there's no depression for water to collect in. Adjust the position of the posts if needed. Let the concrete dry about twenty-four hours.

7. When all the posts are in and the concrete has set, it's time to trim them to an even height.

notches that the building will finally rest. Even where the ground is highest, the perimeter board should not be in contact with it.

FLOOR

Whether you have used concrete blocks or wooden posts for your foundation, the next step is perimeter boards. For sheds up to 12x12 feet, use 2x10 boards; larger than that, use 2x12s.

Before you put the perimeter boards in position on the foundation, nail 2x4 joist supports along what will be the bottom inside edge of the longer boards, as shown in Fig. 24 on page 83 and Fig. 27, below.

Be sure to measure carefully for the joist supports before nailing them on, so they don't foul the posts (if any) or the adjoining perimeter boards.

If you wish, you may use metal joist supports, ready formed and available from your lumberyard, but it's just as easy to make your own.

If you have a concrete-block foundation, simply lay the perimeter boards on the blocks, leaving about 3 inches of block protruding outside the board, as shown in Fig. 24 on page 83. Then nail the perimeter boards together.

If you have a wooden post foundation, set the perimeter boards on the shoulders of the notches already cut. If you were just to nail the boards to the sides of the posts, weight transfer would all be through the nails. But if the post is notched as shown, all stress will be directly on the wood, as it should be. Doublecheck the level; trim the notches if needed. Put the perimeter boards in place and nail them together.

Square the building so far. Do that by measuring corner-to-corner diagonally. If your building is square, the two diagonal measurements should be identical. If you find things out of square, you will probably need a friend to help you push the more distant corners closer.

When the corners are square, hold them that way by nailing in place two temporary bracing boards, as shown in the concrete-block foundation, Fig. 24.

For the rest of the floor laying, follow the directions given earlier in this chapter for putting a wooden floor in a pole shed. Do not, however, extend the floor beyond the perimeter boards.

FRAMING

Walls come next. If you remember three essential things about walls, you will not go far wrong in building them.

First, vertical wall members, or studs, are always set on 16-inch centers unless the space requirements of windows or doors

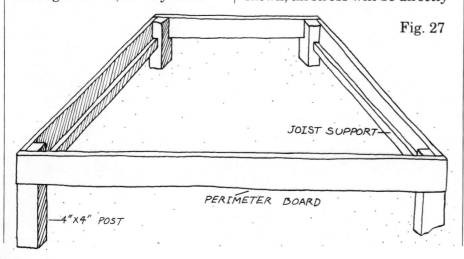

Fig. 27

JOIST SUPPORT

PERIMETER BOARD

4"x4" POST

make that impossible, in which case extra studs are inserted. This is done both for strength and because 4-foot sheets of siding fit neatly with an edge on every third stud. For the latter reason, corner studs are set not on 16-inch centers, but fully within 16 inches of the width of a stud extending beyond the siding.

Second, horizontal wall members—plates, window and door headers, window sills—are nearly always doubled. The one exception to this rule is that the shoe—the long horizontal member at the bottom of the wall—is not doubled, particularly when the structure is a small one.

For illustration of these points, see Fig. 28.

Third, note that the top plates of the front and side walls interlock. The rear wall would have the same pattern as the front wall, but, presumably, without a door. A window could be fitted or not, as wished.

Build the walls as "prefabricated" units, with rough openings framed in for all doors and windows. Use the floor as a work area, building with only a single top plate. Put them in place as they are finished, adding the interlocking second top plate.

When the walls are in place, check that they are all square, both within themselves and with their adjoining walls. If any of the walls are not square, force them until they are, and then nail on temporary bracing boards at 45° angles to the floor and studs. Two supports per wall are sufficient. These braces, by the

way, should also be nailed to the walls you found perfectly square, so that they don't depart from square when you set on the rafters.

For instructions on windows, doors, siding, roof, and finishing touches, see the section on pole-frame sheds earlier in this chapter.

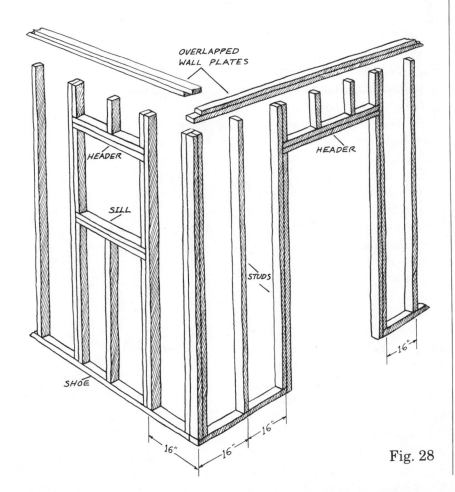

Fig. 28

Hand-Hewn Log Shed

If you live in an area where logs are easily available, they may well be the obvious choice for an unobvious storage shed. Building with logs is strenuous, but not hopelessly so if the building you have in mind isn't large, or if you are sensible about the way you plan to put the logs together.

Log buildings, from open-fronted sheds to six bedroom houses with three-car garages, can be bought ready-cut from several suppliers. Vermont Log Buildings, Hartland, Vermont 05048, and Green Mountain Cabins, Box 190, Chester, Vermont 05143 are two.

Here is a design for a home-built 8x16-foot shed made of logs. The same method of construction can be used for buildings of any size, and has the advantage that no matter how large the structure, no log need be longer than 8 feet. This is no small advantage. First, it means the logs can all be handled by one reasonably sturdy man—a great advantage over traditional log construction. Second, it means logs can be used that would be unsuitable for conventional log building, whether because of lack of straightness and length, or because of too many branches. It is a lot easier to find a reasonably straight and clear piece of log eight feet long than it is to find one, say, twice that long. Third, in areas where pulp or plywood logging is taking place, logs about 8 feet long are easy to

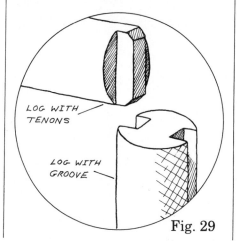

LOG WITH TENONS

LOG WITH GROOVE

Fig. 29

get—even to have delivered to your building site—because that is the size to which pulp or plywood loggers cut them for ease of handling and trucking.

Unless you are in such an area, or have some other nearby ready supply of straight (which usually means coniferous) logs, log building is probably not for you. You really need to pick and choose the logs on the spot, and then if you have to have them hauled a long distance to your site, it will cost a bundle of money.

Nor should you try log construction unless you have either great skill with a chain saw or a lot of time. One or the other is essential. All the tenons and grooves can be cut with a chain saw—if you know how—or they can be cut with a hand saw and hammer and chisel—if you have the time. But don't try to learn advanced chain-saw techniques on a project like this, unless you have plenty of time to practice, and lots of extra logs to practice on.

Although I have seen an Indian man build a cabin using this technique with green logs, I'd strongly recommend you give the logs a year to season, preferably stacked in ricks off the ground so air can circulate

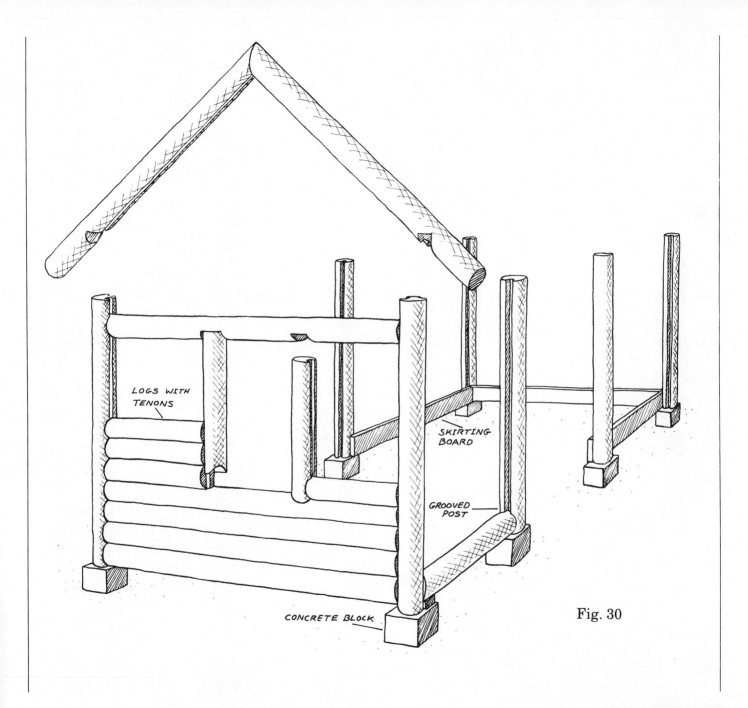

LOGS WITH
TENONS

SKIRTING
BOARD

GROOVED
POST

CONCRETE BLOCK

Fig. 30

around them. That will get most of the shrinking out of their system, and after a year most of the bark will peel off easily by hand. You don't have to peel the logs, but it's a good idea, as various wood-boring insects are almost sure to get between the bark and the wood and, in time, so loosen the bark that it will fall off anyway. This will not happen with cedar or redwood, to be sure, but it will with just about any kind of pine. Besides, it's easier to draw your cutting lines on the bare wood than it is on bark. If you decide not to peel off the bark, use a carpenter's chalk line to make your lines with. In fact, a chalk line is handy for marking even a peeled log.

You can, of course, treat the logs with one of the chemical products sold to prevent wood rot and insect infestation, or buy logs that have been commercially treated, but I don't advise it. All chemical treatments smell, and will give the shed an odor you may find unpleasant. Besides, there are plenty of log cabins in the United States and Canada, even ones made of soft pine, that have withstood both rot and insects for a century or more.

The drawing shows how logs are put together in this kind of construction (see Fig. 29, page 88). At each corner, intermittently along each wall (roughly every 8 feet), at door posts, and at window posts are vertical logs, which are grooved lengthwise. The horizontal logs are cut with tenons or tongues at either end, which fit into the grooves in the vertical posts. Basically, that is all there is to it.

Since logs are not uniform and because so much else depends on the girths of the logs you get, a list of materials isn't possible for this building. Given logs the size shown in the sketches—6 to 8 inches in diameter—the 8x16-foot building will require between sixty and eighty 8-foot logs, depending on size and whether you make all your rafters of logs or substitute 2x6s for the inner rafters. The size of the dimension lumber you use for door and window frames will also depend on the thickness of the logs—for logs 6 inches or more in diameter use 2x6 lumber; for smaller logs use 2x4s. If you can get it, use rough-cut full-dimension lumber; it fits in with the feel of logs better than planed boards.

The length of the log spikes you use will also depend on the girth of the logs, especially of the vertical ones, as you must drive the spikes horizontally (or slightly angled downwards) through the posts into the ends of the horizontal logs, to a depth of about 4 inches. For the logs shown here, 6-inch spikes would be about right.

Dig seven holes, about a foot across and about 9 inches deep, and fill them with concrete. They will be the footings. Screed them well, and smooth the tops level, although not necessarily level with one another. Providing your site is not extremely rough, the time for leveling is when you lay down the first course of horizontal logs.

When the concrete has set, place a concrete block on each of the footings, with the holes up-and-down. Fill the holes with concrete, level with the top of the block.

Stand a post, grooved as shown in the drawing, on each of the blocks. During the building, at least until you have two courses of horizontal logs down and nailed in place, you will have to support the posts with temporary props, say 2x4s nailed to stakes in the ground. The finished log structure is heavy enough not to need to be attached to the footings.

Next cut the horizontal logs to length, and cut the tenons. Measure each log individually for its space; don't try to mass-produce. Because cutting exact fits is difficult in this kind of construction, cut the logs about an inch longer than your measurements tell you you will need. Fit and trim the tenons as needed. If you round off the top and bottom corners of the tenons, the logs will slide into place more easily.

After you lay in the first course of logs, do your leveling. Start with a log that abuts the highest footing. Lever it up at the lower end so that it is level, and prop it thus with a slice from the end of a log. Drive one 6-inch spike into each end, horizontally through the post and into the end of the tenon.

Continue around the building, setting each log at the same height as the first. Prop each log as you did the low end of the first one—with a vertically placed slice of the right thickness from a long end—and spike each log into position from both ends. Where a log already in place on the other side of the post prevents you from driving a spike directly in horizontally, drive it in at the shallowest angle you can, from above the adjoining log.

If the logs are very straight and smooth, you just might get away without dressing them, but generally they won't lie properly against one another unless what are to be their abutting sides are dressed smooth. Traditionally dressing was done with an adz, but it goes more quickly with a chain saw, if you're experienced with one. In dressing, take off as little wood as you can, and where possible, dress the bottom side of the log above to fit the top side of the log below. Water drains better off a log with a rounded top, and good drainage is important to prevent rot.

When you reach the level of the windows (four logs up in my shed), set in the vertical posts for the window rough openings, as shown in the drawing. Cut out a saddle on the bottoms of the window posts so that they fit snugly onto the log immediately below the window. At the top, the window posts are cut flat, and the log above them is notched as shown to take them. Both cuts are needed for good drainage.

To install window and door units, box in the openings with 2x4 or 2x6 boards, according to the size of your logs, and then follow the direction given in the section telling about pole-frame construction, page 77.

When you reach the top of the walls at the sides of the building, and want to cap it with the triangular gable section, you can either cut the progressively shorter horizontal logs with flat ends, angled to conform to the pitch of the roof, and then lay on them and nail in place a 1x6 or 1x8 board, forming an end rafter pair, or you can cut angled tenons as shown in the drawing, sloped to conform to the roof pitch, and then groove the end rafter pairs to take those tenons. Either way it is necessary to spike each (progressively shorter) log to the one below it with two or three spikes. Use spikes about four inches longer than the girth of the logs being nailed. In the case of my shed, 12-inch spikes would do well. For the other rafters you can use either logs or 2x6 boards. See page 68 for further details about roof framing.

For the floor, the easiest approach is to nail a ring of 2x8 boards to the bottom course of logs on the inside, rather like the skirting board described in the pole-frame section. Nail the 2x8s on with log spikes, the longest your logs will accept, set about a foot apart. The spikes will carry the floor, so don't stint them. From there proceed as directed for the wooden floor in the pole-frame shed.

PRIVIES

Consider the Privy

I grew up in a house with flush toilets, so privies have very little nostalgia value for me. Still, on a number of occasions, both as child and man, I have been in situations that required the use of a privy, and I have to admit to some memories of time spent in them that was, if not golden, at least good silver plate. There is something nicely comforting about the isolating semidarkness of a privy on a day when you have had your pocket knife taken away, a bigger boy has broken your favorite toy car, and your pet rabbit has bitten you.

In fact there are lots of good reasons for having a privy. Suppose your work can best be accomplished in solitude, so you have built a studio some ways away from your house. Fine for working, but what do you do when you have to go to the bathroom? You could put in a proper toilet and a septic system, but with the necessary running water and extensive excavations, that would be expensive. The solution might well be a privy. It is cheaper to install, it can be as sanitary as a flush toilet, and it will suit the mood of your work hideaway.

Many vacation homes, too, can best be served by a privy. A privy will not freeze and crack in the winter, as will a toilet bowl full of water, and it will require much less construction work than will a complete septic waste-disposal system.

In some terrains, it is virtually impossible to install a septic system; a privy is your best and possibly only answer. My family and I own a small island in

Privies are still found, even in some suburban backyards. This one is no longer used, but is preserved, like the old pump, as an antiquity.

Carter Smith

a north Canadian lake. It is not much more than a granite outcropping, thinly covered with rooty, stony soil, rarely as deep as 18 inches. Extensive blasting would be required to set in a septic tank and laterals. In such a situation a privy is the only practical solution.

You must, of course, check in with your local health department or building inspector before you start building a privy. Local ordinances and schedules of acceptable practices vary greatly from place to place. What may not only be legal but encouraged in Beatrice, Nebraska, may be flagrantly illegal in Marysville, across the border in Kansas, or for that matter, in the next town north in Nebraska. Some health departments are eager to consider imaginative alternatives to sewage treatment, while others remain opposed to anything more daring than a septic system. You had best find out the stance of your local authorities before starting anything.

In some situations, then, a privy is a good solution, but privies have negative qualities, too. As you find them, they generally smell nasty, are fly-infested in the summer, paralyzingly cold in the winter, and tend to be poorly designed for

A traditional privy made of clapboard with a wooden shingle roof, and a rudimentary concrete–block foundation.

the human form. Most of these discomforts, though, can be avoided with proper ventilation (a vent pipe from the pit to above the roof), chemical treatment of the wastes, insect screening, and insecticides. Even heating can be arranged if required.

The greatest advances in privy technology, however, have been made below the ground. If you have any past experience with outhouses, it has probably been with the traditional pit privy, a wooden closet set over a

hole in the ground, which allows the liquid wastes gradually to leach into the surrounding soil. Until recently, all privies were pit privies. In areas of sparse population, or where there are no underground watercourses or nearby bodies of water to pollute—and if the local department of health has no objections—a pit privy is just fine, and will be the cheapest privy to put up.

Within the last few decades alternatives to the traditional pit privy have been devised in answer to problems of population growth and a growing awareness of the evils of pollution. In our modern world of increasing population densities, it is often necessary for the public welfare that sewage waste be strictly contained. In fact, in many suburban areas of our country, it should be much more strictly contained than it is. In places where septic systems abound, many of them inadequate and old, ground water is commonly polluted. As drinking water in those places often comes from wells, the drillers have to go quite deep before they tap into a water source that is free of fecal pollution. Even at that, there is always the chance that some underground watercourse will at some time feed sewer wastes into the drinking-water supply. The water systems of many American localities are periodically so affected.

The modern vault privy offers an answer to all this. The only differences between a pit privy and a vault privy are to be found below floor level. Unlike the pit privy, which has always been a do-it-yourself response to necessity, the vault privy is a result of the work of engineers. The vault is a waterproof holding tank for the waste. It can be made of reinforced concrete, bought ready-made or poured at the site, or of stainless steel or fiberglass. The vaults can be bought from any firm that sells septic tanks—see your Yellow Pages. They must, to be sure, be cleaned out from time to time, but so must septic tanks. The firms that take that job off your shoulders are also listed in the Yellow Pages under Septic Tanks. The vault privy has the advantage over pit privies—and for that matter septic systems— of leaching no waste whatever into the ground. There is absolutely no pollution

An elegant Victorian privy in Colorado. The brackets were made by a master craftsman, but the eave boards would not be difficult to duplicate.

Carleton Knight, III

of the surrounding ground.

The standard vault privy operates on an *anaerobic* principle—that is, the wastes are allowed to decompose, usually with the help of toilet chemicals, in an oxygen-poor environment. There is no forced ventilation of the vault; only the standard vent pipe as described earlier, and the seat opening, which normally is kept closed. The methane gas produced by the wastes escapes harmlessly through the vent pipe, and comparatively little fresh air enters.

In recent years *aerobic* (oxygen-rich) treatment facilities have come increasingly into use. In them air is forced through the waste, greatly speeding the decomposition process. In such systems some mechanical elements have to be introduced—air pumps or other devices to keep the wastes stirred up. The treated waste is an excellent and bacteriologically safe fertilizer. This kind of system is sensible only if it is supplied with a larger quantity of waste than a family privy normally receives. Electricity is required for the aeration and agitation machinery.

Composting is another method of fecal waste treatment that is receiving some attention.

The pit of a composting privy has two sections, separated by a movable baffle plate. When one side has been used for several months, the baffle plate is moved to the other side so that the waste in the first can ferment and decompose with no fresh waste added to it. When the decomposition process has finished, the waste will have turned into a biologically safe soil that is a superlatively rich fertilizer. (You can tell when the fermentation is finished by taking the temperature of the waste; it will have dropped to that of the surrounding air.) You have only to remove it, either spread it or bag it, and move the baffle plate back again to let the waste in the second side process itself. If

DO-IT-YOURSELF CONCRETE

First of all, figure out the volume of concrete you need in cubic yards. If only a small amount of concrete is required, say less than a quarter cubic yard, it is easiest to buy it premixed with sand and gravel. To prepare it you need only add water and follow the directions on the sack it comes in. A steel wheelbarrow and a spade or flat-bladed garden hoe are all you need for mixing.

If you need between a quarter and a full cubic yard, you can, of course, still use a premixed concrete, but you will save some money if you buy cement, sand, and gravel separately and mix them together yourself. Your hardware or lumber dealer will help you with quantities and the size of gravel best for your project. The usual mix is one shovelful of cement to two shovelfuls of sand to three shovelfuls of gravel. A mixture with more cement and less (or no) gravel will give a smoother surface, but slightly less strength. Add water while mixing until the mixture has the consistency of thick mud.

If you need a full cubic yard or more, you may well find it worth the extra cost to have it delivered by the ready-mix truck. Check with one of the dealers listed under Concrete in the Yellow Pages.

you like gardening, you certainly ought to talk to your health-department people about these systems of producing good fertilizer; see which one they recommend.

If a privy is impossible, either because the ground is too rocky or too swampy, your solution will probably be a chemical toilet. The best ones cost around a hundred dollars, contain their own reusable flushing water, and need to be emptied and refilled with water only about every fifty uses or so. But much less expensive ones can also be bought, which consist of a toilet seat and lid mounted on the top of a can, which in turn has a removable steel liner in which the waste is collected. When the liner is full, it must be carried somewhere and emptied, washed out, and replaced in the can.

This plywood privy cabin has a roof made of plastic sheeting, to let in light. The arched roof has no sharp gable, at which the plastic would eventually begin to tear, but plenty of curvature to shed water and snow. Lattice strips are used to support the plastic, just as they are used to bow out the screening of the tower gazebo (page 220).

Robert Perron

A Basic Pit Privy

Here is a design for a basic pit privy. Even if your needs require one of the more modern privy designs, you will find much in what follows that is still applicable. All privies have some kind of cabin, and all except chemical toilets of the simplest kind require a hole in the ground.

If the terrain and water table permit, you should ideally have a pit about 2x3 feet by 5 feet deep. Very rocky or loose soil may require that you vary those dimensions, but if you can manage it, a pit with a capacity of about 30 cubic feet will last a family from four to twenty years or more, depending on the number of people, whether use is seasonal or year-round, and the percolation rate of the soil.

Under *no* circumstances should the bottom of the pit be less than two feet above the high-water table, that is, the highest level reached by ground water, usually in the spring. If water starts to seep into the pit at any time in the digging, backfill at least 2 feet of soil into the hole, or better—much better—change your mind and go for a water-tight anaerobic vault, or an aerobic or composting system. If you are building your privy in an area with a very high water table, you have no other choice.

FOUNDATION

Unless you have something palatial in mind, the cabin of your privy will be higher than it is wide, which will make it inherently unstable. It was frequently the case that outhouse builders dug a hole and then placed the roughly constructed cabin on top of it, and left it at that. When the pit filled up, they dug a new one and moved the cabin. Even the plans for a privy drawn by the New York State Department of Health propose little more than that. Well, that's fine—until the first strong windstorm, or until rainwater scours away the edge of the hole, allowing the cabin, to settle in.

Much more secure in the

DIGGING A HOLE

If the nature of your soil or your special needs require a wider hole, one you can stand in as you dig, *watch out*! Never dig deeper than your own shoulder height! Most people don't think of digging a hole as a dangerous occupation, but it most certainly can be. A great many people have been killed, suffocated, when one of the walls of the hole caved in. It is entirely appropriate to approach deep-hole digging with a certain amount of healthy fear. Remember: *Never dig deeper than your own shoulder height!*

long run, although more work initially, is a concrete foundation to which the privy is firmly attached. The instructions that follow are for a 6-inch-wide foundation that runs along the perimeter of the privy.

Materials Needed

16 feet of 1x8 board for the concrete form.
eight 6-inch x ¼ inch carriage bolts, with a nut and flat washer each.
25 or 30 2-inch box nails
4 cubic feet of concrete

Construction Steps

1. Dig the pit. Make it as deep as you like—and the water table permits—but bear in mind the caution on page 100: *Don't work in a pit deeper than your shoulder height!* Do not, however, exceed the area dimensions of 2x3 feet; if you do, you will have trouble meeting the specifications for the foundation itself. Area dimensions of 2 feet by 2 feet 8 inches will give you some margin for error here.

2. Cut out a lip 7 inches wide by 6 inches deep around the back and along two sides of the hole.

3. Extend the lipped sides as 7x6-inch trenches to a point 4 feet forward from the rear wall of

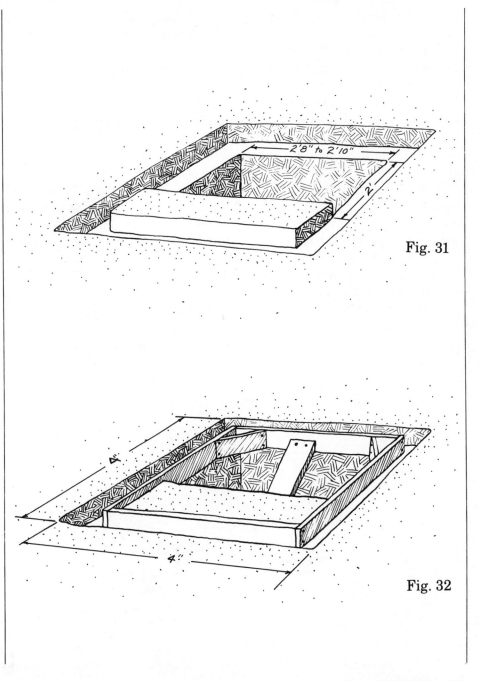

Fig. 31

Fig. 32

the lip, as shown in Fig. 31, on page 101.

4. Dig a 7x6-inch trench along the front to complete the rectangle. This rectangle will form the perimeter of the privy, and should have external dimensions of 4 feet square.

5. Along the inside of the lip erect forms of 1x8 boards, as shown. This will transform the lip into a trench with an earth bottom, one earth side, and one wooden side, Fig. 32, page 101.

6. Use the left-over scraps of 1x8 to brace the corners that are over the hole. It's also a good idea to brace the rear of the form with a prop running diagonally to the bottom of the hole.

7. You now have a 6x6-inch trench all the way around. Pour it full of concrete to ground level. (See the box on page 98 for mixing your own concrete.)

8. Before the concrete hardens, place into it eight 4-inch carriage bolts, heads down, with 2 inches of bolt shank projecting straight out of the concrete, Fig. 33. These bolts should be placed 8 inches in from the corners of the foundation; they should be 2 inches from the outer edge at the front and back, 2½ inches at the sides. They will hold down the 2x4 sills, which will be drilled to receive them. Be sure the bolts are perfectly vertical. Use a square to check.

9. When the concrete is dry (give it 24 hours) remove the board forms by knocking them free with a hammer.

THE PRIVY CABIN

For a traditional cabin, (Fig. 38, page 110) you can't beat a shed-roofed structure. By reason of its purpose and furnishing, you need full standing height only in the front half, so this cabin has a front wall height of 7 feet, sloping to 6 feet at the rear, giving a minimum standing height immediately over the front of the box of a little over 6 feet.

Materials Needed
192 feet of 2x4s, in twelve 16-foot lengths.
five 4x8-foot sheets of ⅝-inch plywood.
24 feet of lath, in three 8-foot pieces, for door stop molding and screen molding.
6 feet of 4-inch stovepipe
damper unit for stovepipe
cap for stovepipe
toilet seat ring and lid
3½-inch (16d) common nails for framing
1¾-inch (5d) common nails for siding, flooring, and roof.
piece of linoleum, 2 feet 6 inches by 4 feet
pair of 4-inch T-hinges with

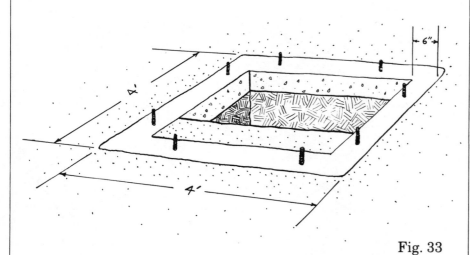

Fig. 33

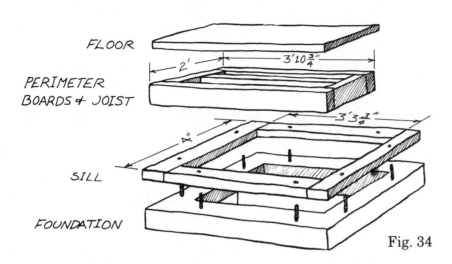

FLOOR

PERIMETER
BOARDS & JOIST

SILL

FOUNDATION

Fig. 34

screws to attach.
thumb latch with screws to
    attach
2 or 3-inch hook and eye
1 square yard of window
screening (requirements
    will vary)
toilet paper holder with screws
    to attach
2-inch duct tape

Construction Steps
2x4s measuring 1½x3½ are
assumed throughout. If yours
vary, you will have to check *all*
length specifications and adjust
as necessary. Use 3½-inch nails
for all framing steps, unless
others are specified.
    1. Cut the sills:
    Two sills, 4 feet
    Two sills, 3 feet 3¾ inches

Lay one of the long sills in
place on what is to be a side wall
of the privy, so that it rests on the
two vertical bolts. Its outer edge
should be ⅝ inch in from the
outer edges of the foundation.
Hold the sill carefully in place,
and hit it with a hammer once,
hard, over each of the two bolts.
Drill ⅜-inch holes exactly at each
dimple left by a bolt. The holes
specified are larger than the
¼-inch necessary, to allow for
some error and adjustment.
Repeat for the other long sill.
Then do the short sills, when the
long ones are loosely in place.
Check that the resultant sill
rectangle measures 4 feet by 3
feet 10¾ inches. Adjust it if
necessary, by enlarging two or
four bolt holes.

2. Build the floor frame.
Cut:
Two perimeter pieces, 3 feet
10¾ inches
Two perimeter pieces, 1 foot 9
inches
One joist, 3 feet 7¾ inches

Nail the five boards together, as
shown in Fig. 34, using 3½ inch
nails. The perimeter should rest
on the sills, just outside the bolts.
If it appears the joist is going to
foul the bolts, move it as required
to either side, and then nail it in
place.
    3. Using a few 1¾-inch
nails, temporarily toenail the
floor frame to the sills, remove
the sills from the foundation
(only three sill pieces are
affected), turn the sill/floor
frame assembly upside down, and
with 3½-inch nails set no more
than a foot apart, nail the sills to
the floor frame. Remove the tack
nails, replace the unit on the
foundation, and once again bolt it
tightly in place.
    4. Cut the floor piece out,
by taking the end 2 feet off one of
the plywood sheets. (Measure
carefully and cut down the
middle of the line, as you will
need the remaining 4x6-foot
piece for the rear wall.) Trim the
end of your floor so it fits the floor
frame—presumably 2 feet by 3

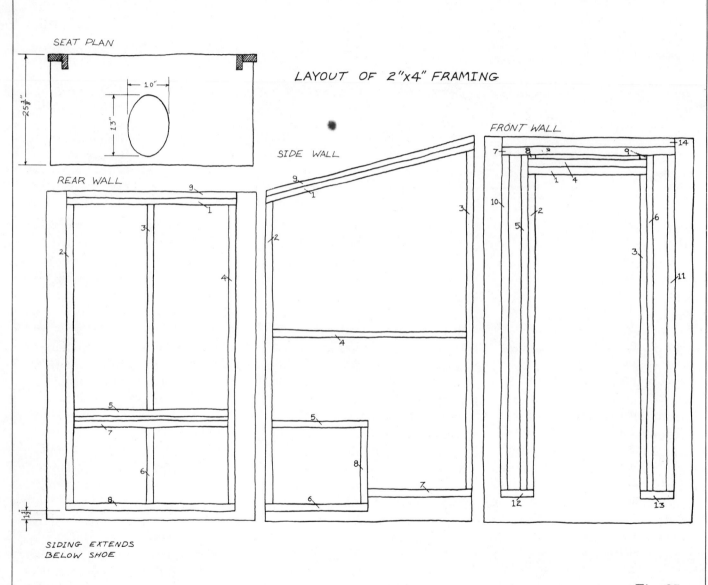

SEAT PLAN

$25\frac{1}{8}$"

10"

13"

LAYOUT OF 2"x4" FRAMING

SIDE WALL

FRONT WALL

REAR WALL

7 8 9 9 14
1 4
10
5 2
6
3
11
12 13

9
1
3
2
3
4
2
4
5
8
7
6

9
1
3
2
4
5
7
6
8

$1\frac{1}{2}$"

SIDING EXTENDS
BELOW SHOE

Fig. 35

feet 10¾ inches. Nail the floor in place with 1¾-inch nails.

5. Now it is time to prefabricate the wall frames. Build the back wall frame first. Cut the following 2x4s:

Two studs, 5 feet 6 inches
One stud, 3 feet 10⅞ inches
One stud, 1 foot 3½ inches
Two plates, 3 feet 3¾ inches
One shoe, 3 feet 3¾ inches
Two medial braces, 3 feet ¾ inch

Assemble the pieces, following the numbered order of the framing diagram Fig. 35 on page 104. Note that stud section 6 must be nailed to the seat support 7 before 7 is nailed between studs 2 and 4. The crucial dimension is ⅝ inch (for the plywood seat) between braces 5 and 7. When the assembly is done, set it aside.

6. For the front wall cut the following 2x4s:

Four studs, 6 feet 2½ inches
Two studs, 5 feet 10 inches
Two plates, 3 feet 3¾ inches
Two door headers, 2 feet 3 inches
Two shoes, 7⅞ inches
Two stud bits, about 1½ inches. Wait to cut these until you can measure accurately for them.

Assemble the pieces, following the numbered order of the framing diagram. Note that the door frame should be assembled before being nailed in place between the studs 5 and 6 and plate 7. Then measure for, cut, and fit the stud bits 8 and 9. If they fit snugly they do not need to be nailed in place. When the assembly is done, set it aside.

7. Except for being mirror images of one another, the side walls are identical. The quantities listed below are for both side walls. Cut these 2x4s:

Two studs, 6 feet 2½ inches, beveled
Two studs, 5 feet 6 inches, beveled
Four plates, 4 feet 2 inches, beveled
Two nailers, 3 feet 9 inches
Four shoes, 2 feet
Two vertical seat supports, 1 foot 3½-inches
Two horizontal seat supports, 1 foot 10½ inches

Note that these four studs must be bevel-cut at one end, and the 4 plates must be bevel-cut at both ends. Here is a simple way to make those cuts accurately: Cut a piece of plain (not corrugated) cardboard exactly to cover the plan of the side wall in Fig. 35 and you will have a gauge for marking the angle cuts. Note also that you can cut one short stud, one long stud, and one plate from one 16-foot 2x4, which means you can make all four of those angle cuts with two passes of an angled saw.

Assemble the pieces, following the numbered order of the framing diagram.

8. Nail the side wall frames to the half floor and the sills. You'll have to drill a hole in the shoe to accommodate the bolt and nut protruding from the sills. When nailing to the sills, angle the nails about 30° in order to avoid hitting the concrete foundation—or, just for this bit, use some 2½- or 2¾-inch nails.

9. Nail the front and back wall frames, first to the floor (in front) and the sill (in back, drilling there for the bolts, too). Note the caution in step 8 above about nailing into the sill. Second, nail through the corner studs of the front and back walls into the corner studs of the side walls. At the front, because of limited hammer-swinging room, you will have to drive the nails in at an angle.

You may have noticed that some features of this frame construction depart in apparently significant ways from the canons of house-frame construction laid down in the Garden Shed chapter. For example, corner studs are not tripled, and the rule calling for a stud at least every 16 inches is flaunted in three of the four walls. In a building this small you can get away with things like that, especially if you are planning to use a siding as heavy and rigid as ⅝-inch plywood. In any case, wind and snow loadings on this building will be small.

10. Time now to make the seat bottom. You'll have maximum light and elbow room if you do the interior carpentry before nailing up the siding. Cut from one end of a fresh sheet of plywood a piece 25⅛ inches wide. (Put the rest of the sheet aside for use later as the roof.) Trim one side down so the piece is 25⅛

inches by 3 feet 10¾ inches. Cutouts must be made to fit the configuration of the rear corner studs. An easy way to do this is to use a small scrap of 2x4 as a pattern. Drawing around it, make lines for a cutout as shown in the seat-plan diagram, and cut accordingly. The lip of the back edge of the seat must slide in between the two horizontal 2x4s on the back wall. Don't nail it in place yet, though. When located as it should be, a 1⅛-inch lip should protrude forward of the horizontal seat support on the side wall.

11. While the seat bottom is still loosely in place, draw in the cut lines for the hole. Don't make it too small, as many outhouse builders do. A good size is an oval 10x13 inches, the long dimension

front to back. The front of the hole should be 1½ inches from the front edge of the plywood. Here is an easy way to mark an oval hole this size: Find the center of the hole by measuring from side to side, and 8 inches from the front edge. Now locate two points: one exactly four inches directly in front of the center point, the other exactly four inches in back of the center point. Into each of these points drive a 1¾-inch nail, just deeply enough that they stand firmly. Mark the front, rear, and side extremities of the oval. Tie a loop of string, just big enough to embrace the two nails and a pencil standing on one of the points you have marked as a side point of the oval. Allowing the string to move freely on both the nails and the pencil, move the pencil around the nails, as far from them as the string will allow. You should have a 10x13-inch oval. Remove the nails and cut out the oval with a power jigsaw or a hand keyhole saw (see box).

At one of the back corners of the seat bottom, cut a 4-inch circular hole. The hole's center should be 6 inches both from the back and the side edge.

The seat, by the way, will be 15 inches from the floor. The

STARTING A CUT

To start a cut in the middle of a board, if you are using a hand saw, you must first drill a hole, with a brace and bit, then work with a keyhole saw until you have a slot long enough for a regular hand saw to fit into. Using either a portable circular saw or a power jigsaw, rest the saw on the front edge of its soleplate, holding it so that the saw blade is just over and just clear of the line on which you want to saw. Then start the saw and slowly tip it back, allowing the blade gradually to bite into the wood and work its way through.

seats in most privies are 2 to 3 inches too high off the floor. This unfortunate characteristic, I suspect, results from ignorance. Thomas Crapper, an Englishman who invented the modern flush toilet*, did considerable research on the question of seat height, and to this day his specification of 15 inches is widely regarded as a standard. Builders of outhouses, however, are seldom plumbers or sanitary engineers, and so are not privy to Mr. Crapper's findings. If they take a standard at all, it is likely to be that of an ordinary straight wooden chair, the seats of which usually come to about 18 inches.

12. Completely finish the seat bottom before putting it back in place and nailing it in. Smooth the edges of the hole with sandpaper, and round off and smooth the edges of the front lip, first with a rasp, then sandpaper. When you rasp a plywood edge, move the rasp only downwards, away from you. If you pull it up while it is still touching the

*It is purely and delightfully coincidental that Mr. Crapper's name incorporates a common slang expression for excrement, although of course it may have had an influence in directing him into his life's work. The word *crap* has been around since the Middle Ages, and comes from the Dutch.

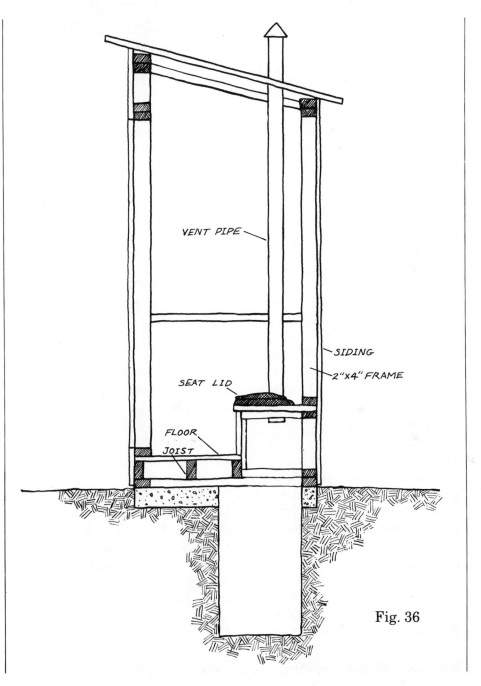

Fig. 36

wood, it will tear up a lot of splinters. As a final precaution against splinters, run a strip of 2-inch duct tape along the front edge of the plywood. Then paint the whole piece, top, bottom and edges. Finally, set it in place, and nail it in. A couple or three 1¾-inch nails at either side is enough. Be careful the support boards of the side wall frame aren't knocked out of place as you nail into them.

13. From the piece of plywood from which you cut the seat bottom, cut another 4-foot-wide strip, 1 foot 2⅜ inches long. (That should leave you a piece of plywood about 4 feet wide and 4 feet 8½ long. Just what you'll need for the roof.) Nail the small piece in place as the front panel of the seat.

14. Time to put on the siding. Start with the side pieces. Hold or tack-nail them in place, and scribe a line, using the top edge of the wall plate as a guide, to indicate where the cut is to be made. Take the side pieces down and cut them, then put them back up and nail them firmly in place.

15. The front siding is a little more complicated, as the door opening must be cut out of it before it is nailed up. The easiest way to do the marking is to tack-nail the front piece in place, then climb inside from the back. You can then mark both the door opening and the place for the top cut, which is ⅞ inch above the top of the upper plate. Cut accordingly, and nail in place.

16. Next nail on the back siding. The 6-foot-long piece left after cutting the floor should just fit, just overlapping the siding on both sides.

17. Now for the roof. From 2x4 lumber cut the following:

Two rafters, 3 feet 5 inches
Four wedges, ⅞ inch high, with bases 3½x1½ inches.

18. Toenail the rafters between the side walls. Set them so their narrow edge is canted at the angle of the roof, and is even with the tops of the side walls. They should be about 16 inches from the front and back wall, and from each other.

19. Nail the wedges in behind the protruding upper lip of the front siding, so that their hypotenuses correspond with the slope of the roof. Use 1¾-inch nails, just two for each wedge. Be careful not to split them. Nail on the piece you have been saving for the roof. Use a chalk line to help you find the rafters and plates when you're nailing. The sides will just overlap the side wall siding. Leave a rear overhang of about 2 inches, and a front overhang of about 5 inches. Don't worry that these overhangs are not supported by rafters. As the roof is strong ⅝-inch plywood, the overhangs don't need any support.

20. Now use the cut-out from the opening to make the door. You can use it just as it is, if you like, or you can edge it with 1x4 boards on the inside, which will make it a little sturdier and look nicer. Remember, though, that the bottom edging board will have to be about 5¾ inches up from the bottom of the door panel, in order not to foul the floor, which the door overlaps by that amount.

21. If you want to cut a vent hole in the door, this is the time to do it. A quarter moon is traditional, but you may fancy some nice scrollwork. See the section on scrollwork on page 191.

22. Decide now whether you want the hinges on the left or right, and hang the door. You will make the job a little easier if you tack-nail the bottom skirt of the door to the floor perimeter board while you screw the hinges on. Leave a gap about the thickness of a shirt cardboard along the hinge side. At the top, the gap

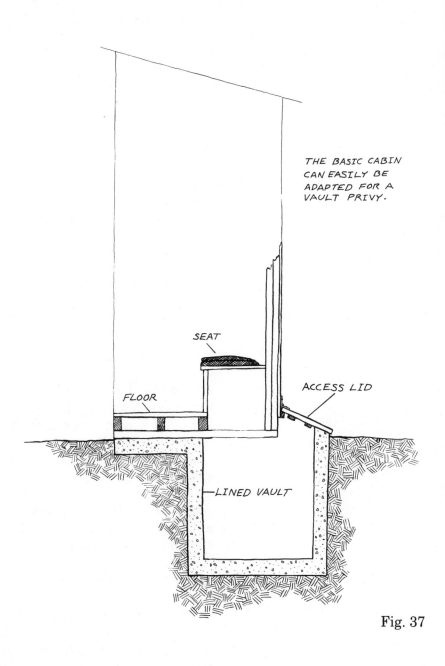

THE BASIC CABIN
CAN EASILY BE
ADAPTED FOR A
VAULT PRIVY.

SEAT

ACCESS LID

FLOOR

LINED VAULT

Fig. 37

from the saw will be just about right. Pull out the tack nails.

23. Install the thumb latch, following the instructions that come with it, and the hook and eye.

24. Time to move inside again, to fit the vent pipe. Put two of the pipe sections together, with the damper section at the bottom, and fit the base into the 4-inch hole in the seat bottom. The damper, by the way, is only really important if you plan to use the privy in cold weather. Everything slows down in the winter, including the chemical processes going on in the pit, so less ventilation is necessary. If you shut the damper you will greatly reduce cold drafts coming up through the hole. Drafts like that can be miserably uncomfortable. You should, however, take the little spring valve out of the small hole in the damper butterfly plate and leave the hole uncovered. The pit should never be completely closed off.

25. Mark for and cut a hole in the roof for the vent pipe. To allow for the pipe passing through it at an angle, you will have to make the hole somewhat oval, with the long dimension front to back.

26. It's time now to consider

Fig. 38

roofing alternatives. Exterior plywood well painted with two or three coats of oil-based enamel will stand up quite well to weather for five to ten years or more, depending more on sun intensity than rain or snow. If you want a longer trouble-free period, you will have to cover the roof with a roofing material. Unfortunately, roofing is usually available only in quantities to cover several times the area of this roof. Still, a roll of tarpaper or a single bundle of shingles, whether wood or composition, isn't expensive, and in time you'll doubtless find a use for the excess. If you use shingles, start with a double thickness at the bottom of the roof, and end with a double thickness at the top. Let both top and bottom rows overlap the plywood by about ½ inch. If you use tarpaper, tuck it around the edges and hold it in place from underneath with lath strips and ¾-inch roofing nails.

27. You can return to the vent pipe now. Pass the third pipe section through from above, and mate it with the sections already in place inside. Fit the flashing collar, and nail it in place. A bead of caulking compound underneath the outer edge of the collar before you nail it down will help keep rain from getting through around the pipe. It is a good idea to press a square of window screening over the top of the pipe before you slide the rain cap on, to keep the bugs out.

28. Now, standing inside the privy with the door shut, nail strips of lath along the studs on both sides of the door, and on the header above, so that the door rests against the edges of the lath when it is latched. This, of course, is the stop molding. If the door is a little warped, don't try to straighten it with the stop molding; attach the molding to conform to the door as it is. It may look a little drunken, but it will do a better job of excluding the flies and mosquitos. A staple gun with $7/16$-or $1/2$-inch staples is handy for fastening the lath strips.

29. Back the vent opening in the door with insect screening, also held in place with lath strips.

30. Time to fit the toilet seat ring and lid. Use a commercial unit for this—they're more comfortable than anything you can make yourself, splinter-free, and quite inexpensive. You'll have to knock off the rubber pads and bend or reset the hinges so that both the ring and the lid lie flat and tight. You want as little odor as possible coming up from the pit.

31. Now it's time to check that you have the privy fly- and mosquito-proof. First the obvious: Be sure the door vent and any other intentional apertures are *tightly* screened. Then, preferably on a bright day, after you have had a chance to accustom yourself to the semidarkness of the privy, look around for small unintentional openings, like cracks between boards or joints that didn't quite meet. Fill any you find with putty or bathtub caulking. Don't think little spaces like that are too small to worry about, by the way; mosquitos have a genius for finding—and getting through—amazingly tiny holes.

Some bugs will get in, of course, if only when you go in and out the door. If you find them troublesome, there are several things you can do. First, before you sit down, spray a good bug killer into the hole. (Now that aerosol cans are being made with a propellant other than Freon, you can spray with a clear conscience, as far as the upper ozone layer is concerned.) Second, hang a sticky fly-paper strip in one of the back corners—well away from where your head might touch it. Third, hang a fly swatter within easy reach. Swatting the bugs at least makes

CHEMICAL TREATMENT OF WASTES

Until recently the standard toilet chemical was chlorinated lime, a white acrid-smelling powder. Now, however, vastly superior toilet chemicals are available that work faster than chlorinated lime and have no offensive odor of their own, if used as directed. A small amount of the chemical poured into the hole breaks down the solid waste in a matter of hours, killing the bacteria and germs, and neutralizing the odors. But be careful to follow the directions on the container. If you use too much chemical, your privy will reek of it, which while less unpleasant than untreated waste or chlorinated lime, still isn't very nice.

The term "toilet chemical" is generic. Hardware and general-store clerks in areas where privies are used will know what you mean.

you feel as though you're winning the war.

32. Paint and otherwise decorate as you like. Whether you do the interior or not is entirely a matter of taste—except for the seat, which you will already have painted. It is nearly impossible to wash off an unpainted wooden surface, and you must be able to wash off the seat. Outside, you must paint the roof, if you haven't used a roofing material, or in a few years you may find the plywood beginning to separate. Take care to get the edges thoroughly impregnated with paint. You will also extend the life of the building if you paint the walls, although paint there isn't so critical as paint on the roof.

33. Cut the piece of linoleum to fit the floor. Leave it loose, so you can remove it for washing.

34. Finally, while sitting on the toilet, find a convenient spot for the toilet-paper holder, and mount it in place. You're all done, at least with the basic article. Electric lights, heat, magazine racks, and so forth you can install as you need or want them.

PLAYHOUSES

How Playhouses Are Used

Being self-centered as most of us are, we aren't much good at divining the wishes and needs of others. All too often we assume they are the same as our own, or at best, what our own would be if we were someone else. This myopic little foible has led to all sorts of vexations. In particular, as far as this chapter is concerned, it has led to a lot of underused playthings and to a lot of huffy resentment on the part of parents, aunts, uncles, and others, of the lukewarm enthusiasm with which their carefully selected gifts are received by the kids.

All children are notorious for selecting as their favorite a scruffy, neon-orange, stuffed rat sent as an afterthought by a distant relative, and for ignoring the expensive, carefully selected, and obviously much more appealing bears and monkeys presented by closer kin. The reasons must remain obscure, but I have a nagging feeling that at least some of this perversity relates to the energy, emotional and otherwise, with which a toy is presented to a child. Therefore, I propose Lane's Law of Successful Toy Giving: *Don't Push*. This goes for playhouses too.

If a playhouse is to be successful, it must have a lot of scope for different kinds of play. There is play which centers around strenuous physical activity. It may involve activity for its own sake, like sliding down a slippery slide or bobbing on a teeter-totter, or it may have the excercise of physical skill as a central ingredient—climbing, throwing a ball, jumping rope. There is play which is a means of establishing contact with playmates. This kind of play often seems aimless or stultifyingly repetitive to grownups, but that's only because they fail to see that it is only a vehicle for companionable togetherness. Then there is play based on fantasy. This element is strong in hide-and-seek and cops-and-robbers, or in board games like Monopoly, but it probably achieves its purest form in plain daydreaming.

A really successful playhouse is conducive to all three kinds of play. For physical activity, and if possible, the exercise of a physical skill, a ladder and a rope or two can fill the need, or, particularly for smaller children, a slide. Where reasonably possible, too, the playhouse should be arranged so that two or more children can play in it together. A playhouse of my childhood was particularly good in this way, for inside it there were two ladders, each leading to a separate trapdoor in the roof. It even had two front doors. In fact, that was about all it had, but my best friend and I didn't mind, and we spent a lot of time in it. It's much more fun for children to be able to play with such things if they can do so simultaneously, and don't have to take turns.

Many garden and home magazines feature plans for backyard structures like this free-form playhouse, which incorporates a useful storage area.

Building scope for fantasy into a playhouse is at once the easiest and hardest job. My best friend and I had no trouble working out an endless series of fantasies with that old double-doored, double-laddered shed, even though it distinctly lacked exotic features. Specialized toys impose rules, and no group of the population, even FBI agents,

is more rule-conscious than little kids. The rules say that if you want to play with a toy firehouse, you have to play fireman, because *that's what it's for.*

The moral, of course, is to make a playhouse that limits the imagination as little as possible. The temptation may be great to make it look like a space rocket or a ship's deck—particularly if the

kid is going through a space jag or you're deeply into sailing and want your son or daughter to be too. But resist it. An amorphous, unindentifiable, sort of Rorschach-inkblot kind of playhouse is almost certain to be more successful in the long run.

DESIGN ELEMENTS FOR A PLAYHOUSE

So, first, a playhouse should allow for open-ended fantasy of all kinds. It should offer some arrangements for physical play, for companionable play, and for imaginative play. It should have open places, private places, high places, and low places. But remember, a feature that invites rough physical exercise may also, under different circumstances, fire the imagination or provide secret and quiet enjoyment.

Take windows for example. Windows in the playhouse should definitely be left unglazed. Glass breaks easily and cuts deeply. And they should be low enough for kids to climb through. On the other hand, windows should be equipped with shutters, preferably ones that can be closed from the inside. They will serve to keep out both rain and prying eyes.

There are several advantages to building the playhouse under the limbs of a large tree. For one thing, the tree will provide shelter during summer rains, frequently enough (if it has a thick foliage) to allow use of the playhouse even on moderately rainy days. For another, you can suspend ropes from a limb of the tree so that they hang just free of the side of the playhouse. Make gaps in the roof railing (if you decide to use one, as described on page 128 so kids can use the ropes to escape either into or out of the playhouse if pursued by bad guys, or they can swing on them at other, more peaceful moments. Hanging ropes can fit into a lot of different fantasies (quite apart from vigilante executions), and they provide a means to plenty of strenuous physical play.

Be sure, though, to get the right kind of rope. Best is heavy natural fiber (hemp or sisal) rope no less than 1 inch, no more than 1½ inches in diameter. Your hardware store probably won't stock rope that heavy, but they can order it. Rope like that is also

Why building inspectors—and parents—turn gray. The vertical posts of this two-story playhouse are dangerously inadequate. The ladder is risky, too, the rungs are too close to the building's side for a good foothold, and they don't look solidly attached.

Authenticated News International

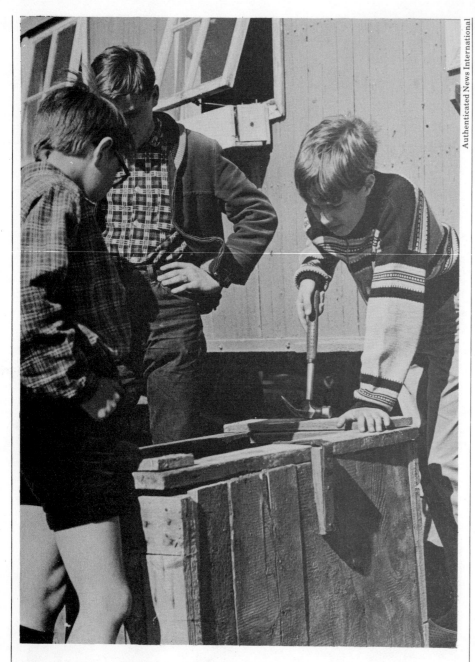

available from a ship chandler. Other places to try are tent and awning stores and rigging companies. Don't be attracted by ropes made from nylon or other artificial fibers. They're stronger for a given diameter than hemp or sisal, but they're slippery, and you specifically *don't* want thin rope. You need something with enough body to it for a child's hands and knees to grip firmly. A quarter-inch nylon rope will hold up a piano, but it's too thin and smooth to grasp properly.

An outside ladder to the roof is also a good thing. Any interior ladder will most likely be vertical, but the outside one can slant a little, making it a bit easier for smaller children to manage. Space the rungs, by the way, rather closer together than you would if you were building a ladder for adults.

A sandbox inside the playhouse, preferably sunk right into the floor, encourages convivial play and it can be used even on rainy days. But don't put the sandbox directly under a roof trap-door so that children can jump into it from above. First,

When you have an interesting building project it's usually easy to get all the help you'll need.

some kid in a moment of forgetfulness is likely to jump down among younger kids playing in the sand, and bloody someone's nose. Second, toys, like a steamshovel with a long, pointy boom, frequently are left in sandboxes, and then covered with sand.

If you have fairly young children, a slippery slide from the roof of the playhouse down to the ground is appreciated. You can build a slide by forming sheet metal over 2x4s nailed on either side of a 1x12, but better yet, buy a commercially made slide and fit it to the playhouse. The manufactured ones usually have dips and humps in them that make them more fun to slide on. Don't be tempted to put a sandbox at the end of the slippery slide for the same safety reasons that you wouldn't put a sandbox under the trapdoor.

There must be at least one secret compartment somewhere; better, a secret compartment for each child you have in mind when you build a playhouse. These compartments don't have to be large—a few cubic inches is enough. If you decide on frame construction, they can go under a loose floorboard with a boxed-in space underneath, You will probably also have little spaces

The essential form of a geodesic dome is determined by the placement of the skeleton framework. A wooden or metal gusset is used to join the five timbers at each joint.

above the walls, below the eaves. They can be used as well. Other possible locations will occur to you as you build.

There should also be a good hiding place, if possible big enough for two kids. Make it separate from the main interior space of the playhouse. A buried oil drum is perfect and will easily accommodate two children at the same time. For instructions on how to install such a feature safely, see box on page 135.

STRATEGIES FOR DESIGN AND CONSTRUCTION

You can't just buckle down and build a playhouse the way you would a gazebo or a doghouse. Oh you can, I suppose, but it's a grave mistake. In the first place, you have to consider who's going to do the designing and building, you or your kids. There's something to be gained either way.

You can look on the playhouse project as a teaching and learning exercise. Let the kids help draw the design, show them how to figure the amount of materials needed, let them discover how hard it is to saw a straight line or drive a straight nail. The problem with this approach, however, is that the project might lose its novelty—and consequently its attraction—after about twenty minutes of actual building on the site. If you manage to keep the kids at it until the playhouse is built, it's a fair bet they'll never want to have anything to do with it again. But if in the process they have learned to wield a hammer and saw with some skill, it won't be a total loss by any means.

You can, of course, let them help in the building only when they feel inclined, and it may well be that that's the way to go about it. You must know by this time pretty well how your kids tick and which approach will work best. But if your children are anything like my son, the best way to handle the whole project is in veils of deepest mystery. If I announce to my son that I'm about to build something like a playhouse, he immediately is wildly enthusiastic and comes up with zillions of neat ideas for features it ought to have—*must* have. That's all fine, of course—creativity and imagination are great things but I'm not in a position to provide any but the smallest fraction of the spiffy features he's dreamed up. I simply can't afford to build him a seventeen-room playhouse on five floors with an infinitude of electric eyes and burglar-proof secret compartments. But that's what he's set up in his imagination, so the finished reality is doomed from the outset. A playhouse gets a poor start in life if it's anticlimactic.

TEMPORARY OR PERMANENT?

How long can you reasonably hope or expect this playhouse to be used? One summer? Two? Four? No more than four, and more likely fewer, if you're thinking of fairly intensive use. That's for one child, mind you. If you have twenty or so children like J.S. Bach, the playhouse can look forward to almost indefinite use. The first grandchildren will be coming into season just as the last of the first generation are outgrowing it. But if each of your children gets even one good solid season of enjoyment out of the playhouse, the effort will have been worthwhile.

In recognition of this inescapable truth, there are two courses of action you can take. You can either build a playhouse out of cheap materials that will last only one season and can be easily disposed of, or you can build a playhouse that can serve some different purpose once it has been abandoned by its original tenants.

There's something to be said for both approaches. Using cardboard appliance cartons, for example, it is easily possible to build a splendid playhouse that will last a summer at least, and cost next to nothing to build. If it doesn't last more than one season, and if your kids want another playhouse next summer, then you can get some more appliance cartons and make another one. That way you have the charm of annual novelty and nothing more to dispose of in the fall than some large boxes.

Another type of one-season playhouse to consider is the common sapling-and-tarpaulin kind described in most scouting and camping manuals, or the wattle-and-daub dome and the baled-hay house, both described later in this chapter.

Temporary Playhouse From Cardboard Cartons

Here are suggestions for a playhouse made out of discarded appliance cartons. I say "suggestions" rather than "plans" because the various elements of the playhouse could be put together in a great many different ways, and much will depend on the cartons you are able to get.

Half the fun in this project is in collecting the boxes, and then working out the most interesting way to put them together. There is almost infinite scope for imagination. The only important caution you need to keep in mind throughout is to plan your cuts in such a way as to weaken the cartons as little as possible. And remember, by gluing them together on a stable foundation, you increase their strength.

Before you do anything with any cartons, paint them well inside and out with two, better three, coats of an exterior oil-based enamel. One thing unprotected cardboard doesn't take well is getting wet, and as water is bound to get inside this playhouse, the inside must be as well protected as the outside.

Don't by the way, underestimate the strength of these cardboard cartons. Cardboard gets a bad rap for being flimsy. The heavy stuff they make appliance cartons of isn't that at all. Three or four years ago I made a "privacy booth" for my wife's third-grade classroom from a refrigerator carton with a door cut in one side and a small, high window on the other. No bracing of any kind was added. Even after years of hard schoolroom use, the box is still in fine shape, although it needs repainting.

When I decided to build an appliance-carton playhouse, I went down to our local dealer to see what he had on hand in the way of cartons. There for the taking were three refrigerator cartons and one built-in oven carton. Two of the refrigerator cartons measured 33x33x70 inches, the third was 30x39x70, and the oven carton was 33x32x29. Now 33 inches and 33 inches and 30 inches make 96 inches, exactly 8 feet, so that gave me an idea for a foundation and a roof. In the cellar I happened to have a 4x8-foot sheet of ¼-inch hardboard and two 4x8-foot sheets of ⅜-inch exterior plywood. Just right. A rigid roof is a helpful strengthener too. It both distributes the load of children playing on top and helps prevent twisting and deformation of the cardboard.

So on a level piece of ground under a large tulip tree I laid

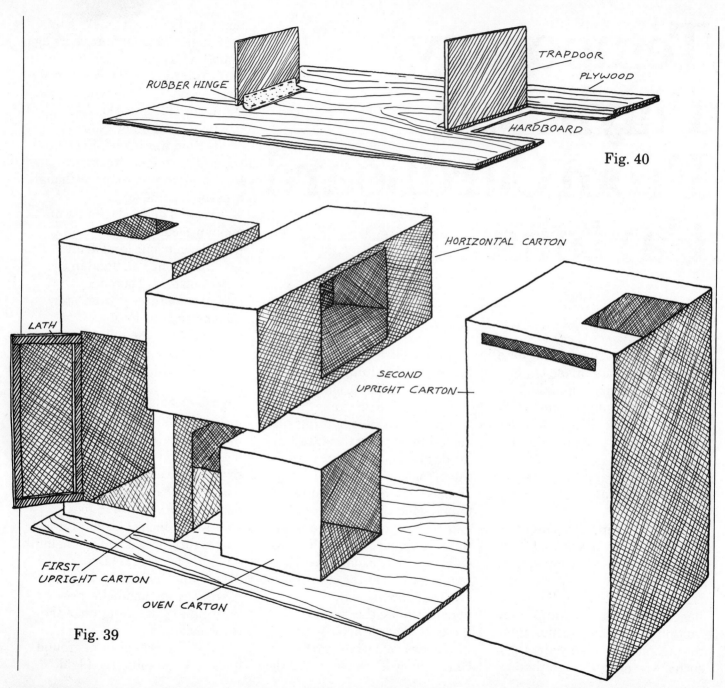

RUBBER HINGE

TRAPDOOR

PLYWOOD

HARDBOARD

Fig. 40

HORIZONTAL CARTON

LATH

SECOND
UPRIGHT CARTON

FIRST
UPRIGHT CARTON

OVEN CARTON

Fig. 39

down one of the sheets of plywood. Then I took one of the 33x33x70 refrigerator cartons, its paint dry by now, and cut a 2x4-foot opening in one side for a door, saving the cutout to install later as the door. I made the bottom of the door opening 6 inches above the bottom of the carton so as not to reduce its strength any more than I could help. A flat kitchen knife made a good cutter, but I had to sharpen it afterward. Heavy cardboard is tough stuff and dulls any knife quickly.

Using linoleum glue, I glued the carton upright on one end of the sheet of plywood. (It's the one on the left in Fig. 39.) Then, right down at the bottom of the side of the carton facing the rest of the plywood sheet, I cut a hole 31 inches wide by 28 inches high.

Next, I cut the top and bottom off the 33-inch-high oven carton and glued it, lying on its side, right next to the refrigerator carton, so that the open top of the oven carton just fit against the hole I had cut in the side of the first carton.

At the bottom of the 39-inch side of the 30x39x70 refrigerator carton I cut another 31x28-inch hole, and just 2 inches above it, still another hole, 31 inches wide by 32 inches high (not shown in drawing).

I then glued the second refrigerator carton upright to the plywood, mating the 31x28-inch hole up to the remaining open end of the oven carton.

I could now enter the door in the first upright carton, get onto my hands and knees, and crawl through the oven carton to the other upright carton.

Then, right in the middle of one side of the remaining 33x33x70-inch refrigerator carton, I cut a hole 32 inches wide by 31 inches high. Then I laid the carton on its side, on top of the oven carton, between the two upright cartons, so that the hole I had just cut mated up with the upper of the two holes in the second refrigerator carton, and glued it in place. It was now possible to crawl through the oven carton into the second upright carton, stand up, and climb into the horizontal refrigerator carton, and lie down. Except for the roof (with trap-doors) and a few minor matters, the construction was done.

In the center of each end of the hardboard sheet I then cut a 22x22-inch notch, and glued the sheet to the tops of the two upright cartons, bridging them.

Since the 33-inch width of the first carton, plus 33-inch length of the oven carton, plus the 30-inch width of the second upright carton came, all together, to exactly 8 feet, the sheet of hardboard just spanned the two refrigerator cartons. Then, working from a step ladder, I cut holes in the tops of the upright refrigerator cartons exactly matching the notches in each end of the hardboard.

Finally, I cut 24x23-inch notches from the ends of the remaining piece of plywood, and glued it to the hardboard so that the notches in the plywood were centered exactly over those in the hardboard, exposing just an inch of hardboard on each of the three interior sides of the notch.

Using a 3-inch-wide strip of inner-tube for a hinge, fastened in place with a staple gun, I placed the plywood cutouts back into the notches, and voilà, trapdoors! The notches in the hardboard were cut an inch smaller on all sides to give the trapdoor something to seat against, see Fig. 40.

An old wooden 10-foot ladder I had was just right, cut in half, to provide access to the trapdoors, inside the upright cartons. I fastened it to the sides of the cartons with U-bolts. Where the ends of the U-bolts went through to the outside of the cartons, I sandwiched 6-inch-

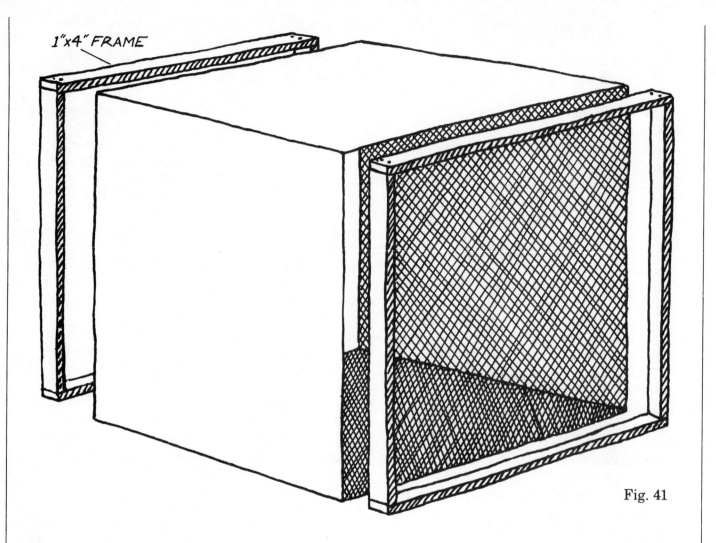

Fig. 41

square pieces of hardboard against the outside of the card-board, so as to better distribute the pressure of the U-bolts against the cardboard.

After finishing with all that, I braced the bottom of the

interior opening into the horizontal refrigerator carton, above the oven carton. As an additional brace, I made two square frames of 1x4s to fit into both of the cut-out ends of the oven carton, see Fig. 41. A

framework of ¼x2-inch lath around the inside of the door and the inside of the door frame might be a good thing as an extra precaution, see Fig. 42.

Several finishing-up jobs. The hinge for the front door I

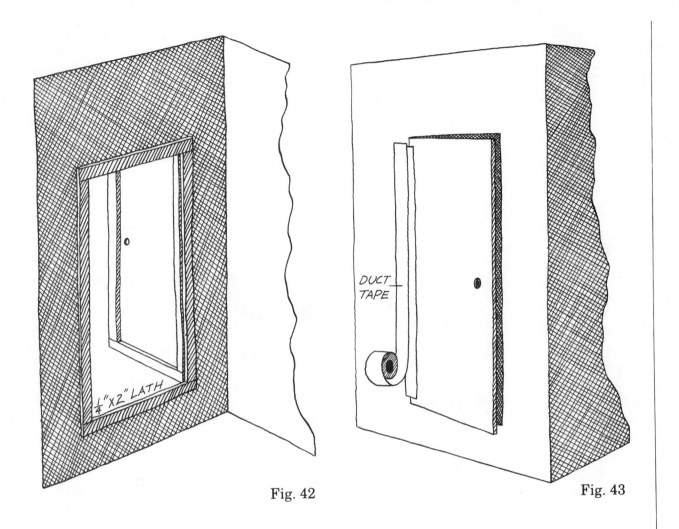

Fig. 42

Fig. 43

made out of a double strip of duct tape, a very sticky, very strong, cloth-backed tape, 2 inches wide, originally marketed to seal leaks in hot-air heating systems, but extremely useful for all kinds of other jobs, see Fig. 43. All hardware stores have it. Some kind of latch openable from the inside and outside is necessary; it will be easy to mount because the door has a light wooden framework. Check with your local hardware store as to what types of latches will work for you; I made my own with a sliding piece of 1x2. Then I cut slits here and there for ventilation, light, and looking out. I made them about 15 inches long and 2 inches wide. Cut like that, they do not significantly weaken the cardboard, and are narrow enough not to let in much rain. In any case, I only cut them high in the sides of the upright cartons, under the overhanging eaves of

Finally, in order to exclude water, I stuck masking tape over all the exposed edges where I cut the cardboard and asked the kids to be careful to leave it in place. If water gets inside the cardboard, everything will fall apart in days.

And last of all, a surprise. I found myself left with a little open breezeway slot between the roof and the upper side of the horizontal refrigerator carton. It occurred to me to close off the ends of that "breezeway" and cut a couple of access holes to the space in the upper side of the horizontal refrigerator carton. Secret compartments!

It's likely, of course, that if you decide to build an appliance-carton playhouse, you won't come up with cartons quite the same size as those I found. As it is, the cartons I got could have been put together in lots of different ways. This is one project for which there is no "right way."

Fig. 44

A Permanent Playhouse

If you like the idea of building something permanent, a basic shed playhouse—a good kind, since imagination can turn it into anything—can be built in the same general way as a good garden shed, such as those in Chapter Three. Later, it can double as a summer house or gazebo or even a garden shed when the kids are bored with it.

Pole-frame construction is particularly good for a playhouse; with the poles set well into the ground, even a small structure is very stable, and no foundation is required.

If you decide on house-frame construction, then you must anchor the structure well to the ground, particularly if it is small, less than 8x10 feet. A number of children all leaning over the roof edge on one side could topple a small, unanchored playhouse, with unfortunate consequences.

ANCHOR FOR HOUSE-FRAME CONSTRUCTION

Materials Needed
8 concrete blocks, 8x8x16
one 100-pound bag of premixed cement
four 20-inch long, ⅜-inch diameter threaded rods
4 each, ⅜-inch nuts and flat washers

Construction Steps
1. Dig four holes under what are to be the corners of the playhouse, about 10 by 18 inches in area and a foot deep.
2. Lay two blocks in each hole, with holes running vertically, one on top of the other. The top block should protrude about 4 inches above the ground.
3. Mix up the concrete and with it fill up the holes in the blocks. If you have any left, pour it into the holes around the outside of the blocks.
4. Screed the concrete—that is, work out the air bubbles—with the threaded rods, by moving them rapidly up and down in the block holes for a minute or two. Then set them upright (use a level) in the center of one of the block holes, one rod to each pair of blocks. Use the middle hole if your blocks have three holes; either if the blocks have only two holes.
5. When the concrete has set firm (give it 4 or 5 hours at least), tamp earth into the holes around the blocks.
6. Drill holes in sills to fit over the vertical rod ends, set the boards in place, lay on the washers, and draw the nuts up tight. Nail the perimeter boards to the sills.

SIDING

Much the easiest siding to use is plywood—⅝-inch for a pole frame. Directions for siding (page 74) and fitting doors and windows (page 75) are given in the chapter on garden sheds.

ROOF

If the roof is to be a play area, support it with 2x6 rafters set on 16-inch centers and cover them with ¾-inch plywood. Half-inch plywood over 2x4 rafters would the plywood-hardboard roof, so unless rain is driven by a high wind, it shouldn't come in at all. support the children safely, but there would be enough give to induce leaking. Slope the roof about 1 inch for each foot of run. That will shed water but still be easy to keep balance on. For ways to make a nonskid surface (important even if you're going to have a railing), see page 143 in the chapter about treehouses.

RAILING

An option to consider is a railing for the roof. Use 2x4s for the balusters (posts) and top rail, and 1x4 for the center rail. The balusters should be between 32 and 48 inches apart; the size of your playhouse will determine the spacing. In the front and rear of the house, nail the balusters to the rafter ends, the flat side of the balusters against the butt end of the rafters. On the sides nail the balusters to the sides of the end rafters. Use two 4½-inch common nails for each joint if you have used 2x4 rafters; use three if you have used 2x6 rafters. In the former case, cut your balusters 28 inches long; in the latter case, 30 inches long. Again using two common nails for each

Playhouses like this can be bought partly assembled from many toy stores and lumber yards. They are inexpensive, last long, and adapt well to play of every kind.

post, nail the 2x4 rail to the tops of the balusters, lying horizontally, wide side down. Cut 45° angles at the ends to allow the side rails to join nicely to the front and back rails. Halfway between the top rail and the roof, nail on the center rail. It should be affixed to the inside of the balusters. Be sure to sand the upper surface and edges of the top rail well, to minimize the danger of splinters. The railing will be 26 inches high, which is just right for children up to 4 feet tall.

Wattle-and-Daub Playhouse

A playhouse is a perfect opportunity to experiment with some of the more unusual building methods. Here is a playhouse, simple to make and surprisingly durable, which faithfully recreates the kind of house our primitive ancestors lived in, whether in Asia, Africa, Europe, or on the American continent. It consists simply of some saplings, something to tie them together with, a lot of twigs, and mud, and it is easy to build. Kids will thoroughly enjoy helping with the building.

Construction Steps

1. Cut a number of saplings, no larger at the base than 1½ inches, and as tall as possible. For a small play hut 5 feet in diameter and 4 to 5 feet high, 15 or 20 saplings will be enough; for a hut 8 feet in diameter and about 7 feet high you will need about twice that number. The kind of sapling is not important, but they must be supple, which means they must be cut when the sap is up, in late spring, summer, or early fall.

2. Trim all twigs and branches off the saplings, but don't throw them away.

3. Using as many saplings as necessary, make a hoop on the ground, the diameter of the hut you wish to build. Tie butt to tip, and overlap about 18 inches. Our ancestors used leather thongs or pieces of vine or natural fibers to tie the saplings together. Today binder twine (available in big balls from hardware of farm-supply stores) is easier to come by and works fine. Wrap the twine around several times for a distance of a few inches before knotting, see Fig. 45.

4. Construct two arches, set at 90° to one another, using 2

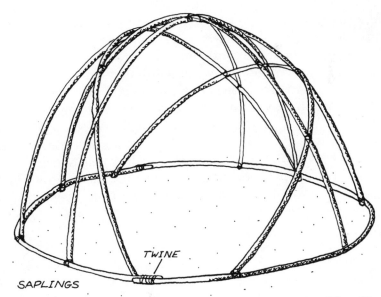

TWINE

SAPLINGS

Fig. 45

DAUB

Daub is basically mud. At its fanciest it is clay mixed with straw. When used by itself, either built up glob by glob into walls, or first formed into bricks and dried, it is called adobe. When used as a plaster and filler on a wall made of woven twigs and saplings, it is called daub. The woven twigs and saplings are called wattle.

Any earth, providing it isn't sandy or rocky, can serve as the raw material for daub or adobe, but earth with a high clay content is better. Pure clay is best. Clay, of course, comes in a huge variety, depending on where you live. The advantages of one clay over another are not great enough to worry about. Use what you get. Scattered through even the richest farmland in the United States and Canada there are extensive clay deposits, so you should be able to find some within a reasonable distance from your home.

The daub is easy to prepare. Best get hold of a supply of decent reasonably pure clay, first of all. As I said, any earth will do, but clay dries harder and erodes less readily than loamy earth. After all, you might as well have the best daub you can get.

Mix the clay with water (a wheelbarrow makes a good mixing trough), adding the water slowly and *mixing* the while until you have something about the consistency of softened, room-temperature butter—not runny, but not really firm, either. Mix in straw until there is enough straw to have permeated the clay thoroughly, but by no means so much that you appear to have more straw than clay. Two good double handfuls of straw in half a wheelbarrow full of clay mixture won't be too far off right. You may have to add some water to keep the butterlike consistency.

You can either gather your own straw (any thoroughly dried long-stem grass will do fine), or you can buy it from a feed store or stable-supplies company. Look under Feed in the Yellow Pages.

saplings for each. Place the butts of the saplings inside the ground hoop and tie them to it, looping and crisscrossing the twine around both. Bend opposite saplings together, then bind them together at the height you want for the house, or that your saplings will permit. Again, overlap by at least 18 inches.

5. Now make and tie in place a number of "great circles," arches extending at various angles, other than vertically, from the ground hoop and following a "great circle route" around what is to be the dome of the hut. As before, tie the butt ends of the saplings to the ground hoop, and if more than one sapling is used for a great circle, overlap the tips about 18 inches. Where great circles intersect with one another or with the two initial arches, tie the intersections together. Continue to make great circles until you no longer have any openings in the structure larger than 2 feet square, other than the door opening.

6. Using the twigs and branches from the saplings, which you have stripped of their leaves, weave shut all the openings through the saplings except those you want to preserve as windows and, of course, the door, see Fig. 46. You will certainly need to gather more twigs and small branches for this job, but those you have already cut from the building saplings will give you a good start. You have finished when there are no openings left in the wattle (that is what the woven "fabric" of twigs is called) large enough for a man to pass his fist through. In any case, the more tightly you weave the wattle, Fig. 47, the stronger the structure will be.

7. Mix up your daub (see box on page 130) and with your hands apply it to the wattle. Kids of all ages will particularly enjoy this part of the job. Everybody involved in the daubing should wear old clothes and plan on a swim or a bath afterwards. The top layer of daub over the wattle should be about an inch or inch and a half thick. If you want the deluxe job, also apply a thin layer of daub to the inside of the wattle, enough just to cover.

8. A piece of tanned hide would be in keeping as a door covering, but failing that a piece of canvas or leatherette would do nicely.

Daub has the disadvantage of being slowly but surely water soluble. If you have smoothed the daub well on the outside it will shed water well, but you will find from time to time, particularly after a heavy rain, that repairs are necessary. Just mix up a little more daub and fill in what the rains have washed away. Alternatively, if you are expecting a real gully-washer or a long siege of wet weather, cover the hut with a tarp. It is particularly advisable to do this if the daub has not had time to dry out thoroughly after being applied. That process will take about a week of warm, dry, sunny days, longer if weather conditions are less than ideal.

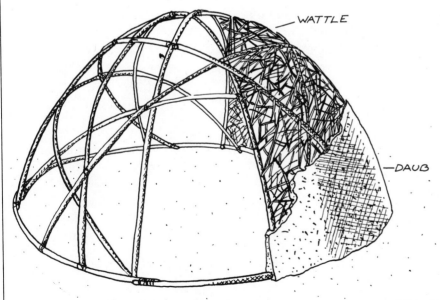

Fig. 46

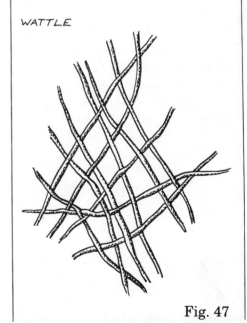

Fig. 47

THATCHING A ROOF

You can thatch with just about any long grass or reed that grows. Of the common grasses, wheat straw is held by some to be the best, but any other cereal grass will work fine too. So will bulrushes, marsh grass, cattail stalks, or tall sedge, as long as the grass or reed is fairly long (at least 2 feet).

Straw can be bought where horse or pet supplies are sold, but it will be much cheaper if you cut the grasses or reeds yourself. If you should be so fortunate as to live within reach of a farmer who doesn't cut his cereal crop with a combine or who cuts his hay with an old-fashioned sickle-bar, and will sell you the unbaled long cut grasses, seize the opportunity.

If you are going to use reeds or reedlike grasses, they must be broken down into their long fibers first. To do this, take hold of a handful of the reeds by their bottom ends and whip them against a post or the top of a fence until they are soft and supple.

The next step (the first step after cutting, if you are using a grass of any kind), is to bind the leaves or fibers into bundles, called switches, that are about 1½ inches in diameter where they are bound tightly at the bottom (ground) end of the plant. For your binding, use binder twine, (available in large balls from hardware and farm-supply stores) as it is strong and resistant to rot. Don't just tie a ring around the switch; wrap it for an inch or so, as though you were making a kind of short handle for a whiskbroom, see Fig. 48.

It doesn't matter whether your material is dry or fresh when you begin laying it on the roof. It must, of course, eventually dry, and the roof itself is as good a drying rack as any. If your switches are about 2 feet long and about an inch and a half in diameter where tied, you will need about 24.

A thatched roof works best and lasts longest if it has a pretty steep pitch. Don't figure on a pitch of less than 45°; if you go to 60°, even better. First, put up an ordinary rafter roof, as you would for any frame structure (see the chapter on garden sheds, pages 68 and 82). Then nail laths across it. horizontally, leaving about an inch space between laths. The laths are what you will be tying the switches to, see Fig. 49.

THATCHING SWITCH

1½"

Fig. 48

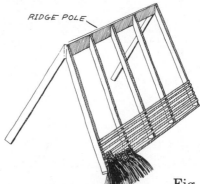

RIDGE POLE

Fig. 49

Tie the switches to the lath with binder twine. Start, just as you would for a shingle roof, at the eave and as you would with shingles, make your first row with two layers. Tie the first double-thick row of switches onto a lath about 4 inches from the edge. This will leave about 16 inches of switch hanging over the edge. The switches should be tied onto the lath touching one another, side to side, but not wedged together. Eight switches to a running foot is about right for each layer. Tie your next row of switches, a single layer now, onto a lath about 4 inches up from the lath to which the first row is tied, and in that manner progress up the roof, one row at a time.

Sealing out the water at the ridge pole (the board at the peak) is the hardest job, and there are several ways of going about it. The least attractive way, but admittedly effective, is to finish off with a strip of metal flashing, wide enough to overlap both sides of the ridge by about 6 inches.

Then there are several ways I know to do the job with thatch. One involves making a lot of special ridge switches, which are tied in the middle rather than at one end. Tie these along the ridge pole, with one end hanging down each side of the roof. These switches should be pressed tightly together, unlike the switches covering the rest of the roof, see Fig. 50.

The second way uses switches of the same kind that you used on the rest of the roof, except that they are tied tightly not just at the end, but are bound for a length of about five inches from the end. Lay these switches alternately on each slope of the roof, as shown. Tie them onto the ridge pole with a stiff bound part extending beyond the ridge, see Fig. 51.

The third way to thatch the ridge requires switches like the ones you made for the rest of the roof, but about twice as thick around. Part these switches, then, in the middle, and drape half over one side of the ridge and half over the other, as shown in the drawing. Tie each switch in place on both sides of the ridge. Press the switches as close together as you can, see Fig. 52.

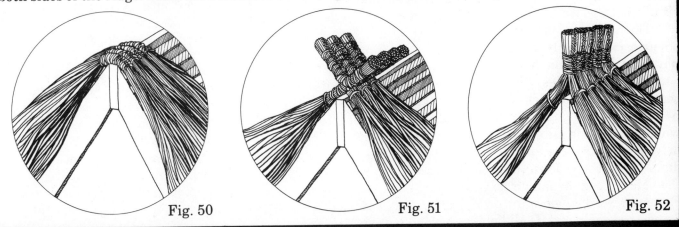

Fig. 50   Fig. 51   Fig. 52

Alternative Playhouses

BALED-HAY PLAYHOUSE

Believe it or not, baled hay is a recognized building material that was used extensively on the Great Plains in the United States and Canada from the late nineteenth century until the Second World War, although lately mainly for temporary animal sheds. But over the years baled hay has been used to build countless houses, schools, and on at least one occasion, quite a sizable barn as well.

The construction method is obvious, particularly if all you have in mind is a playhouse.

Pile up hay bales, staggering them as you would bricks. It helps to hold things steady if, when you're done piling, you drive long sticks (say an inch in diameter at least) down through the bales to peg them together. Fasten in wood window and door frames the same way. Drill holes in the boards and when you have them in place, drive longish sticks into the hay through the holes with a heavy hammer.

For a hay playhouse, I would suggest as a roof a sheet of ½-inch or ⅝-inch plywood, pegged in place like the window and door frames.

If you want to be fancy and preserve your hay-bale structure for a few years (or decades, for that matter), you can plaster it inside and out. That also helps keep down one of the chief drawbacks of hay-bale construction: fleas. Mind you, hay bales won't have any more fleas than did the grass that makes them up, and that may well be few or none at all. But once fleas do get into a hay-bale structure, they are, according to reports of those who know, very difficult to dislodge, since enough insecticide to kill them effectively approaches quantities beyond human tolerances.

LOG-HOUSE PLAYHOUSE

Finally, consider some advantages of a log playhouse. If you have access to logs in your area, a log playhouse wouldn't be hard to build, log structures are solid, long-lasting, and attractive. Such a playhouse should certainly have plenty of potential for fantasy generation. See the Garden Sheds chapter for ideas on building with logs.

ADOBE PLAYHOUSE

There aren't many ecologically purer ways to build than with adobe. It is, after all, nothing but dried mud with straw in it. The most common way to use adobe is in brick form (bricks are usually about 12x6x3 inches, but suit yourself), although it is possible to make your walls by mixing the stuff a bit thicker and slopping it on by the handful, smoothing with your hands as you go. The problem is that the walls dry terribly slowly unless you are in a desert climate, and severe cracking is common.

You make adobe bricks just like regular bricks, except that you don't fire them in a kiln. Let

SUPER HIDING PLACE FOR ANY PLAYHOUSE

Here is a suggestion for a way to arrange a large hiding place, big enough for two kids—providing you do not live in an area with a very high ground water level or very shallow bedrock.

Under the floor of the playhouse, dig a hole large enough to set in a standard 55-gallon steel drum (about 2 feet in diameter and 3 feet tall). Check with your local oil distributor as to where to buy a good drum. Get one with a removable, not a fixed lid.

To install, punch a few holes in the bottom for drainage with a hammer and cold chisel and set the drum in place over about 6 inches of coarse gravel.

Leave about 3 inches of drum protruding above the ground level, and there must also be several inches of space between the top of the drum and the floor of the playhouse.

This makes an excellent hiding place, and because of the air space between the top of the drum and floor level, a safe one.

them dry in the shade about a month.

Form the bricks in molds, open-topped boxes made of 1-inch lumber. The bricks need only stay in the molds until they have become firm—a few hours in warm sun. As a shortage of molds will likely be a bottleneck in your brick production, make a number of them if you can. It helps a lot in getting the formed bricks out of the molds if, before you nail the molds together, you sandpaper the inner surfaces smooth. Then, when the molds are assembled, smear them inside with grease. Alternatively, line the molds with waxed paper.

Good adobe earth may be found almost anywhere, by the way, not only in the Southwest. It became such a popular building material for conventional construction there chiefly because of the lack of wood and because the climate is ideally suited to a building material which must dry quickly and is somewhat water-soluable.

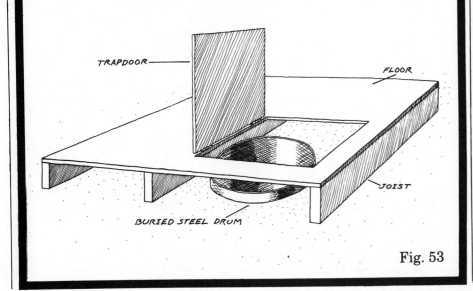

TRAPDOOR

FLOOR

JOIST

BURIED STEEL DRUM

Fig. 53

TREEHOUSES

Treehouses

Of all the little houses in this book, the treehouse offers the most perfect solitude for people who crave it from time to time.

Part of the reason for this is to be found in the close association that treehouses have, naturally, with trees. Trees *may* be nothing but big plants, but that isn't how people throughout history and in every corner of the world have regarded them. Common in many places and times is the belief that trees are living, sensate, perhaps even rational beings. A logger in northern Canada told me once that he could hear trees scream. When he cut one down, he said, he tried to do it as quickly as possible so as not to prolong the agony. "How the hell do we get so smart," he said, "we know what a tree feels and what it don't? Well, I heard 'em screamin', and it ain't the saw, neither. You can hear 'em over the saw."

When the tree itself is not seen as sensately alive, it is often a symbol for life. The great ash tree, Yggdrasil, of Old Norse mythology, is seen as the very primordial source of life, knowledge, fate, even of time and space. Yggdrasil's stem supports the earth, while its branches shelter the whole world and drip honey-dew. Odin hung nine nights on Yggdrasil, offering himself to himself. Some tree.

The origins of the Christmas tree are widely thought to rest in its role as a symbol of life and the regeneration of life, and in a number of societies in Africa and Asia, it is traditional to decorate or otherwise reverence a tree at the time of the vernal equinox. The Freudians have a lot to say about all that.

So trees have a very special

This early 18th-century treehouse was inspired by Defoe's Robinson Crusoe. The trees are probably not structural members, but simply captives of the building.

mystique for us—some of which rubs off on treehouses. Since trees appear benevolent, sheltering, and peaceful, a treehouse transmits all those same feelings—and separation from the world as well.

A treehouse, too, offers us a new outlook on our surroundings. We live, for the most part, in a horizontal world. The natural at-rest sighting of our eyes is more or less parallel to the ground we stand on. Even when we are half reclining in an easy chair or lying in bed, reading, we tend to tip the chin toward the chest, but only so far that our eyes can still comfortably look straight ahead. But get yourself up in a treehouse, alone among the leaves (which you are accustomed to looking up at) and gaze at the familiar ground below (which you are *un*accustomed to looking down on, at least from that vantage point), and the feeling of separation from the familiar world is considerable. Like the fireman hurtling in his truck past red lights and stop signs, you are for a brief while above the law, separated from

A good, basic treehouse with a shingled roof, in a very sturdy-looking tree.

everyday social constraints.

Everyone, I think, should at least once have the opportunity to drive a fire engine and blow the siren, and also to spend several hours in a treehouse. Both experiences are strangely tonic and liberating.

WHAT IS A TREEHOUSE?

Most treehouses, at least as we commonly think of them, aren't houses at all, but platforms, which speaks for the naturally sheltering feeling we have about trees. You simply don't need walls or a roof to get the nice, protected, cocooned feeling that you want from a little house, which the tree itself provides. This assumes, of course, a full, spreading, leafy tree; if all you have in your own yard are tall, spindly, scrawny birches, you may well want walls and a roof.

Of course, a treehouse doesn't by any means have to be built up in the branches of a tree to be a treehouse. The illustrations on these pages show just how loosely you can apply the term. What else, after all, could you call the home of the Turkish shoemaker (see page 143) inside the old plane tree? There can't be much room in that house, but oh what good feelings

it must radiate! Then, too, there are treehouses with trees growing through them, out of them, and even on top of them. Whether you're willing to call them all "treehouses" is up to you. Certainly they would all be different or impossible without trees.

TREE CARE

From the tree's point of view, the best time to build a treehouse is in late fall or winter, when the sap has receded and the tree is dormant. The worst time is spring, when the sap is rising.

The tree at the center of this 19th-century fantasy has been truncated, but there is no reason why it would have to have been. Notice the inverted triangular construction, which must have been very strong.

Any cuts you make then will ooze sap and inhibit the tree's vitality. Unless you need to do a very great deal of pruning, building a treehouse in mid or late summer shouldn't hurt a healthy tree, but winter or summer, any limb or branch stumps more than an inch in diameter should be painted with tree dressing.

Much of a tree's vital juices flow up through the cambium, the layer just under the bark, so if the bark is badly scored more than halfway around the trunk, the tree will likely die. For this reason if you use cables to support your treehouse, protect the trunk or limb around which the cable is passed by threading it through a section of garden hose. If there is wind movement, a bare cable will soon slice through the bark. Furthermore, if there is some part of the tree onto which you must frequently step, and if you are expecting any considerable traffic, protect that spot with a small piece of board nailed in place, onto which people can step instead of stepping directly on the bark. It will not hurt a tree to drive nails or spikes into it.

If you build in a grove or forest, you may want to thin out some of the neighbors of your treehouse tree. It may help your view, and it will help the vitality of the chosen tree, as it will no longer have to share the water and nutrients in the surrounding soil, and may have freer access to sunlight. If you do this, be selective; stunted or sick trees should be the first to go, then trees which by their proximity are hindering the healthy development of the treehouse tree.

If you build on one or more horizontal limbs, you may want to prune off the end of the limb, beyond the part you need to support the treehouse, in order to reduce sway. But be careful. If you cut the limb off too close to the tree, you will kill the limb, and it will eventually (again, after some years) either dry and become brittle, or rot. If, however, you cut it leaving a substantial number of leaves on its twigs and branchlets so that the sap has somewhere to flow, it will live, even if you've lopped off quite a lot of it. If you would like advice, your local county agricultural extension agent will be helpful.

GENERAL NOTES ON TREEHOUSE CONSTRUCTION

When you start to plan a treehouse, forget most of what you've read about the correct distance between floor joists and studs on 16-inch centers, and the like. Above all, don't be wedded to rectangular shapes and square corners. Trees don't conform to building codes (but the treehouse may have to—check with your building inspector), so you will have to arrange your construction to conform to the tree. Some flexibility is preferable to absolute rigidity. You don't want much give, but if there is no possibility at all for some tree movement, the stresses will be unacceptably high and something is going to break or pull loose.

If your treehouse is to be built in a single tree, attached entirely to a thick section of trunk and not resting on a limb or joining two or more trees, you can and should build for rigid strength. The "Greek cross" frame on page 153 is ideal for this situation.

But if you build a treehouse that joins two or more trunks or branches that may move somewhat relative to one another, then you must plan on some compliance between the parts, even if you tie the branches together with a cable and timber brace (see page 155). Some angular twisting, for example, may be expected in a platform. It should not, therefore, be floored with a solid sheet of plywood, as that will tend to resist such movement. Instead it should be floored with 1x4 or 1x6 boards, nailed in place only at their ends, at the perimeter of the floor frame. That will allow the boards to slide gently against one another as the frame flexes.

Compliance in the walls and roof is less likely to be necessary, unless you have built around a tree trunk small enough to flex some in a wind, or unless you have incorporated an upper branch in your roof or some other part of the upper structure. In that case, you can either help stabilize the relative movements with a cable and timber brace, or you can allow for the necessary compliance. Wood will not stretch or compress, but it will bend or twist a little without harm. Standard house-frame construction will allow for some bending and twisting, particularly if you sheathe it with boards, as suggested above for the floor.

The final finishing of the platform floor presents a problem. I built a plywood-floored platform once and brushed it with a couple of coats of the very newest poly-urethane epoxy varnish. It gave a lovely, smooth finish—so smooth, in fact, that when it was wet it had all the surface grip of oiled ice. Treehouse floors will get wet, and you don't want kids—or you—skidding off into space. Here is one solution: A friend tells me he has used

Here is a different—but wonderful—treehouse. This old hollow plain tree was home to a Turkish shoemaker. You can't get much cozier than this.

ordinary, good-quality spar varnish and sprinkled sand on it while it was still wet, with good results. I would worry a little that grains of sand would work loose and leave little holes through which moisture could penetrate, get between the varnish and the wood, and cause the varnish to peel. My friend says no, because you sprinkle the sand on after the varnish has been applied, and while the grains do imbed themselves, they don't penetrate all the way through the coating to the wood, that is, if you give the varnish plenty of time to dry thoroughly before walking on it. Another possible solution to the skid problem might be those nonskid strips that hardware stores sell to put inside your bathtub. Heaven knows, water doesn't seem to hurt them.

STAIRS AND LADDERS

Finally there is the matter of access to the treehouse that is up high in a tree. If you want to make access easy and if the height is no more than 10 or 12

This Mogul miniature painting, "Anwari and Companion Drinking on a Tree-Top", pictures a more elegant treehouse than most you'll see.

feet, build a stairway. If you like, you can hinge the stairs at the upper end so that they can be pulled up, like a fire escape, using a trusty block and tackle, to prevent access by unauthorized persons. Simple stairs consist of stringers (the side boards) and treads (the parts you step on). For heavy duty or if they will need to be more than 12 feet long, use 2x12 stringers. For lighter duty, but strong enough to support the weight of a heavy man, use 1x12s. Lumber 1x10 or 1x12 is adequate for the treads, if they are not to be more than 18 inches wide. If wider, use 2-inch-thick lumber.

Here are directions to build a simple staircase, see Fig. 54, page 146. They are specifically aimed at stairs for a treehouse, but following them you should have no trouble making simple stairs for other applications, too.

1. Measure from the top surface of the treehouse floor to the point on the ground on which you want the stairs to rest. Remember, the steeper you make the stairs, the shorter and lighter they will be—useful if you're planning to hinge them at the top for raising and lowering.

2. Provide a foundation for the stringers to rest on. A brick under each one is entirely sufficient, but if you use two separate foundations, be sure they are level with one another. A small concrete block or a flat stone would do as well. It's best to bury whatever you use at least partly in the ground, to keep it from moving about, or being kicked.

3. Lay one stringer in place and mark it to fit the side of the treehouse and to set squarely on its foundation below. To be sure the horizontal cut at the bottom *is* horizontal, use a carpenter's level as a straight edge for marking the cut line.

4. If your treehouse floor is level and you have been careful to get the stringer foundations level, you can use the first stringer as a pattern for cutting the second.

5. Lay both stringers in place and secure them at the top. Sixteen inches is a good width for such stairs, and allows you to use 1-inch lumber for the treads. The stringers may either be nailed in place (see Fig. 54, page 146) or fixed with hinges, so that the stairs may be drawn up.

6. To find the number of treads you will need, first measure the vertical (not running) height of the stairs, from the floor surface of the treehouse to the ground. If the ground isn't level, place one end of a straight board on the point on the ground to which you have measured, directly below the edge of the treehouse floor, and get someone to support the board over the point at which the stringers rest, and measure the vertical distance from the bottom of the board to the stringer foundation. Make sure with a level that the board is being held horizontally. Add that distance to the ground-treehouse floor distance, and you will have the vertical rise of the stairs. Standard stair treads are 7½ to 8¼ inches or so. When you know how high the staircase must be, and you have decided on the rise you want for each individual step, you can divide rise into height for the number of steps you will need. Remember that your top step is the floor of the treehouse, so you don't need a separate piece of wood for it. But remember that distance in your calculations! You will probably have to deviate somewhat from your ideal rise for each step, in order to have the steps evenly spaced, top to bottom.

7. When you have determined the rise you want for each step, use a carpenter's square to locate the leading top edge of the bottom tread on the

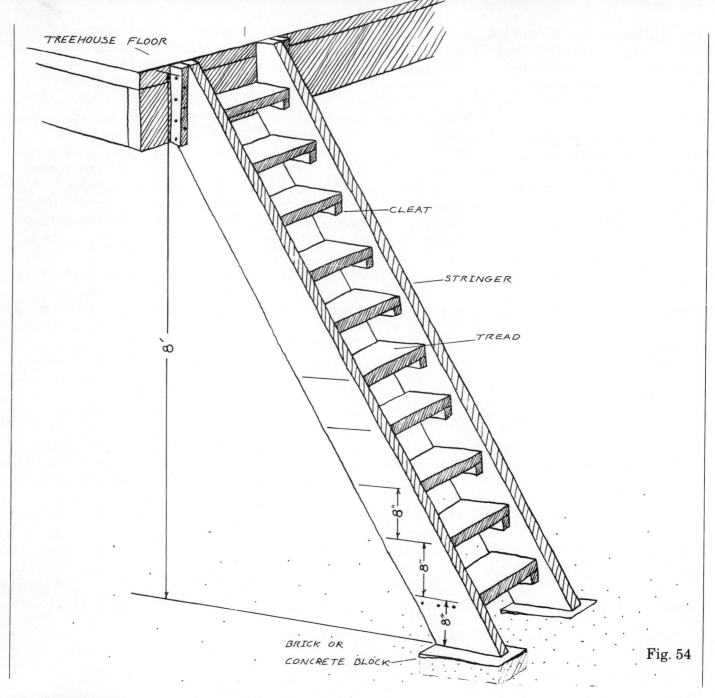

TREEHOUSE FLOOR

CLEAT

STRINGER

TREAD

8'

8"

8"

8"

BRICK OR
CONCRETE BLOCK

Fig. 54

stringers. Let us say you have decided on 8 inches. Place one edge of the square on the horizontal line you have cut at the bottom of the stringer. Then move the square along that line until the 8-inch mark on the vertical scale just meets the edge of the stringer. Make a mark there. Then measure the run along the angled upper edge of the stringer from the base to the mark. Transfer that measurement to the back side (bottom) of the stringer, and make a mark. A horizontal line should join the two marked points, which itself will locate the top of the first stair tread. From this point you can proceed simply by measuring the appropriate distance along the top and bottom edges of the stringer, until you have all the stair treads located.

8. Now carefully measure the distance between the stringers at the top, where they have been fixed into position. If you think you know the measurement, measure again. You need to be accurate to within $1/16$ inch. Be sure the bases of the stringers are just where they are supposed to be before you accept your measurements as final.

9. Cut as many treads as you need to exactly that length. If you have used 1-inch lumber, support each tread with cleats, that is, lengths of 1x2 as long as the tread is deep, nailed to the stringer under each side of the tread, as shown in the drawing. If you have used 2-inch lumber for the treads, the cleats are not necessary, if each tread is nailed in place with at least three $3\frac{1}{2}$-inch nails on each side.

Other commonly used means of access are, in descending order of ease of use, a standard ladder nailed in place at the top, see Fig. 60, page 156, a rope ladder with wooden rungs, a rope with knots at convenient intervals, and a rope without knots. A rope without knots will assure you a lot of privacy in this day of lamentably out-of-condition citizenry.

Best for a rope ladder is $\frac{1}{2}$-inch or $\frac{3}{4}$-inch hemp or sisal rope. It will probably take some trial and error to get the knots in the right places (about a foot apart is good, less for small children), and don't forget to thread a rung (use 2x4s—or 2-inch dowels, if your lumberyard has them) before you tie the knot above it! Attach the rope ladder to the treehouse floor by running it through hitching rings (your hardware store should have them) and back through a second pair of holes in the top rung, and tying knots then in the end. Screw the rings firmly to the treehouse floor, about 6 inches in from the edge. With a round rasp make grooves in the edge where the ropes pass over, so they won't fray so quickly. In any case, check them frequently for fraying, and replace them when it begins.

Best for a simple climbing rope is $1\frac{1}{2}$-inch hemp or sisal. Knots 18 inches apart will make it easier to climb. But be sure to tie a knot in the bottom of the rope, or bind the end with cord, to keep it from raveling. Climbing ropes are also best fastened at the top with hitching rings, but get one just large enough for the rope to pass through easily, and then, after threading it, tie a knot in the top end of the rope to keep it from passing through the ring. Do not tie the rope to itself, as such knots come undone much more easily. And again, check the rope for fraying regularly, and replace it when needed.

Beware, however, the old stand-by of pieces of 2x4 nailed to the trunk of the tree. The least important disadvantage of that arrangement is that a vertical ladder of 2x4s is uncomfortable to climb. More to the point is the fact that it is unsafe. Because of

In some parts of the tropics people have been building treehouses for centuries. This sturdy, well-proportioned, water-tight home shows what can be achieved with practice.

the geometry of straight 2x4s and cylindrical tree trunks, you can nail the boards in place only in the middle. But because of the same geometrical facts, you must step on and hold onto the ends of the boards when you climb. You will be applying a twisting force on the nails, which are located so as to be least able to withstand a twisting force. The result is that the 2x4 will pivot under your weight, your foot or hand will slip off, and you will fall.

Of course you can brace the 2x4s so they can't twist, but that still leaves you with another problem. When you climb a more or less vertical ladder, your hands pull at the 2x4s as you climb, eventually pulling them out of the tree trunk altogether. There is also the possibility that a board will split on you as you have hold of it and send you sprawling. All in all, a makeshift ladder of boards nailed to the tree is a bad idea.

BUILDING A TREEHOUSE

Ultimately, beyond providing a few tips and suggestions, no one can tell you just how to build a treehouse. Partly that's because every tree is different, and each must be adapted to individually. Then too, a treehouse is a very personal thing, and to be satisfactory to the builder, it must express his or her fantasies and whims. In fact, the very environment in which the treehouse is built forces the builder to improvise, again and again.

The drawings and descriptions that follow are merely starting points. You will have to adapt each plan to your individual needs, but there are important ground rules to remember. The points of main importance are: plan well, brace well, and frequently during the building check the integrity of the construction. Beyond that, anything goes. In many cases it will not be possible to adhere closely to standard framing practices, nor should that trouble the builder. No one has any business criticizing your treehouse if a corner isn't square here, or if the roof is a bit crooked there. "I like it that way," is all the response you need make. That, after all, is the way treehouses are—satisfying, forgiving, and entirely self-justifying.

Ground-Level House in a Grove of Trees

If you have a dense grove of trees, you might like to build a ground-level treehouse, using the trees as a frame. If your grove is large, you may want to choose a group of trees well inside the grove to support the treehouse, leaving plenty of trees around it for privacy. Be sure to plan for wind movement in the building of this house. Here is one way to allow for flexing of the frame as the trees move in the wind.

To begin with, construct a perimeter frame of 2x6s or 2x8s nailed to the bases of the trees, as shown in Fig. 55a, a few inches above ground level. Make corners by notching each board, as shown, and fitting them together without nails. The result is sturdy but yet has some give.

Because the side of the tree is hardly ever vertical, nail a wedge-shaped piece of wood between each perimeter board and the tree trunk. Again, refer to the illustration, Fig. 55b.

You are now ready to lay the floor joists inside the perimeter boards. If the house is going to be more or less rectangular, you can try laying them in the normal way, on 16- or 24-inch centers.

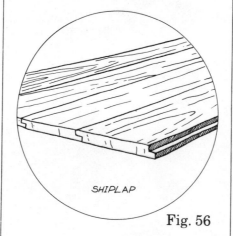

SHIPLAP

Fig. 56

Bevel the ends of the joists so that they meet the perimeter board angle properly. (If the plan of the house turns out to be more nearly circular, then some kind of radial joists, or a combination of parallel and radial joists, will probably work out best.)

For flooring, use shiplap (see Fig. 56) or ordinary 1-inch boards. Nail the boards only at their ends but not to each other, to permit needed compliance.

If you decide you want walls and a roof, then attach another set of perimeter boards to the trees, with their top edges where you have decided ceiling height should be.

Apply vertical 1x4 or 1x6 siding boards either tongue-and-groove or shiplap. The boards that butt up against the trees will have to be sawn to match as nearly as you can the contours of the tree. Like the floor boards, the siding boards should be nailed to the perimeter boards top and bottom; they should not be nailed to one another. Attached this way, they will have a slight—but sufficient—freedom to flex and slip against one another as the trunks move. For chinking between the boards and the trees, use a non-hardening glazing putty.

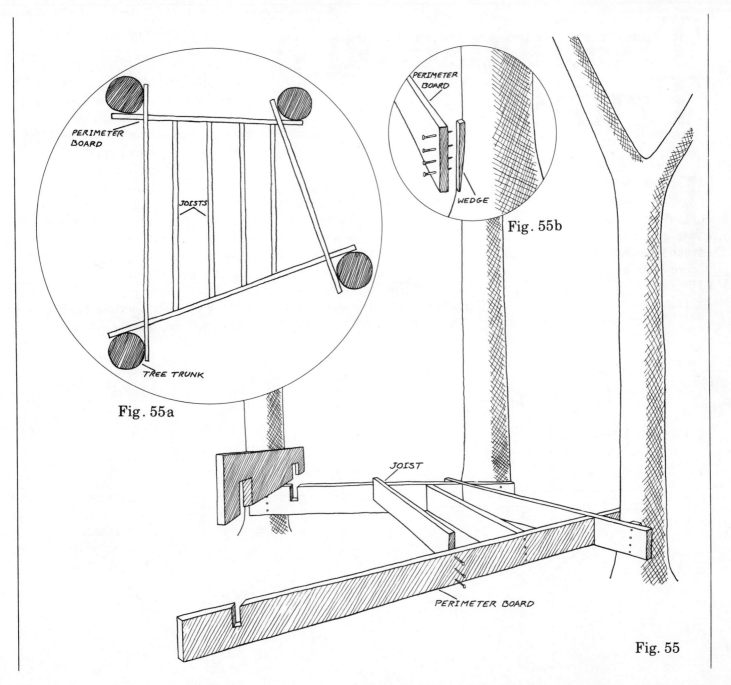

PERIMETER
BOARD

JOISTS

TREE TRUNK

Fig. 55a

PERIMETER
BOARD

WEDGE

Fig. 55b

JOIST

PERIMETER BOARD

Fig. 55

Treehouse on a Trunk

It is perfectly possible to build a treehouse around or attached to the trunk of a tree, and to nothing else. A treehouse like that can be built at any height you find suitable. Here is a suggestion for such an arrangement.

The four main supports are 16-foot 2x10s. For the perimeter boards and floor joists, 2x6s are stout enough. The braces can be 2x4s. There must be a brace for each end of each of the 16-footers—eight braces in all. Note that the boards are *not* notched but nailed together; in this version of the treehouse no compliance is needed, see Figs. 57a and b.

Nail the first two 16-footers, one on each side of the tree, and parallel to one another. Nail the next two 16-footers at right angles to the first two and resting on them. In order to bring the level of the two lower 16-footers up to the level of the two top 16-footers, lay additional lengths of 2x10s on the top edge of the lower boards. These should be secured to the boards not by butt-nailing but by 10-inch wide strips of ⅜-inch plywood running along the joint, as shown in Fig.58. For the floor, use ⅝-inch plywood.

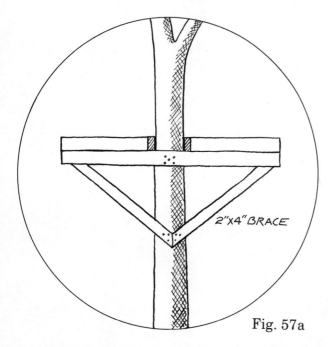

Fig. 57a

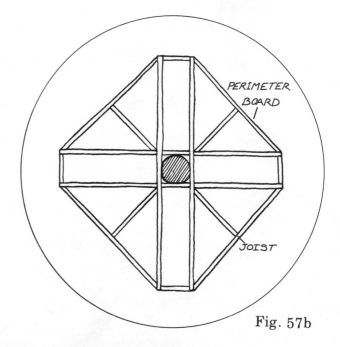

Fig. 57b

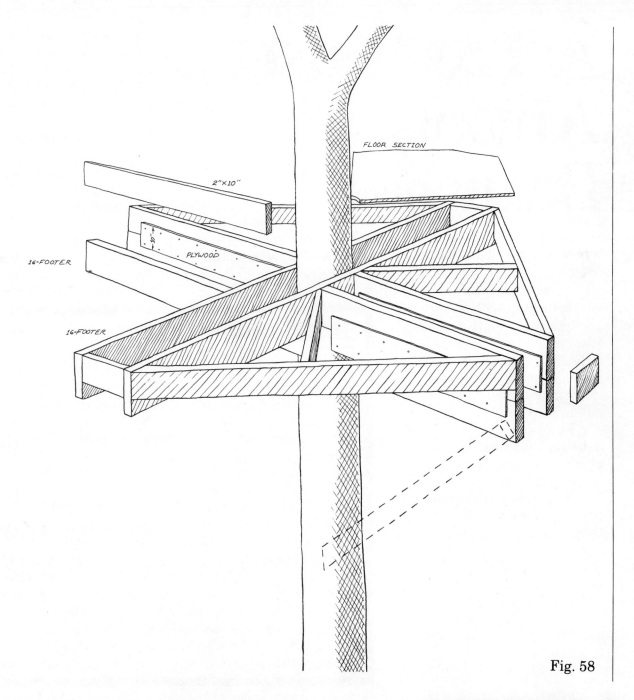

FLOOR SECTION

2"×10"

16-FOOTER

PLYWOOD

10"

16-FOOTER

Fig. 58

A Treehouse Among the Branches

For most people, though, a standard elevated treehouse—with or without walls and roof—seems just fine, and rightly so. Because every tree or set of trees is different, I'm not going to offer a set of plans, but there are important ground rules about building.

First, and most important, don't get your treehouse all planned and then go looking for a tree to put it in.

Find the tree first, climb up to the level where you want the treehouse, sit down on a limb, and spend a nice long time thinking and staring at what the tree itself has to offer you. Keep at it until you begin to see the treehouse taking shape on, in, and around the tree and its branches.

It is, of course, nice if you can build on outstretching limbs. For many trees, however, this implies a fairly substantial elevation, one that may well be too high for little kids to feel comfortable about. It is wise to keep their feelings about height in mind when you are building, and a perfectly good treehouse can be built supported by the trunk above.

There is no reason why you must build an elevated treehouse in just one tree. A treehouse can be built between two trees. There are, unfortunately, some problems with using more than one tree, since two trees standing next to one another in a

BLOCK AND TACKLE

The only real pain in the neck about building an elevated treehouse is getting the lumber up to the building site. To be sure, if the treehouse is to be a fairly modest affair and not higher off the ground than you can conveniently reach with a stepladder, the problem will not be acute. But if any of the boards are heavier than you can easily hold in place with one hand from a ladder, while with the other hand you drive nails in, or if you plan to build beyond the reach of your ladders, then you simply *must* rig up a block and tackle. To be sure, that means reaching a fairly sturdy limb *above* the place where you plan to build, but that is what you must do. Not having seen your tree, I can't suggest how best to do that. So hook or bind the block (pulley) to that higher limb and thread it with the tackle (rope). Tie a knot in the rope to prevent it from running back out of the pulley. Then, at a convenient height at the base of the tree, drive in a good stout spike, to which you can secure the rope after hauling a board up. The block and tackle will provide you with a fine "sky hook" from which to hand your lumber as you are nailing it into place.

windstorm do not necessarily wave about in unison. This isn't hard to deal with, though. If you bind the two trees together with a wire cable and brace them apart with timber, as shown in Fig. 59, you should have no difficulty with out-of-phase movement. A good way to do that is to cut a 4x4 timber so that it fits snugly between the two trees. Toenail it into place. Then run a cable between the two trees, with a turnbuckle in the middle. Protect the trees from the cable by running it through lengths of garden hose as it passes around them. Tighten the cable well with the turnbuckle. The trees will still wave, but they'll wave together, and you can safely use both of them to support your treehouse.

A nice treehouse can be built across two limbs. Don't worry that the two limbs aren't level with each other. You can deal with the leveling by building up with 2x4s laid flat on the lower limb and piled to the desired height, as shown in Figs. 60 and 60 a or by running posts down from the perimeter boards to the lower limb. (You can also step the floor of the treehouse. A split-level treehouse is very impressive when built, and exudes a feeling of competence

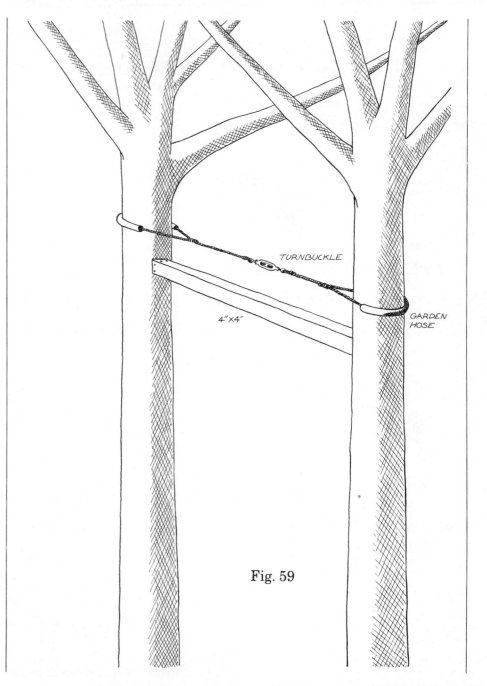

Fig. 59

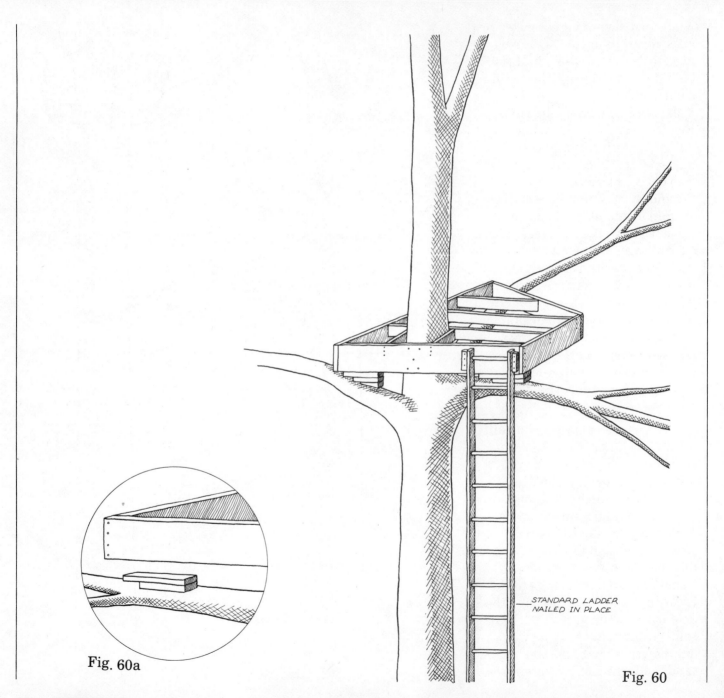

STANDARD LADDER
NAILED IN PLACE

Fig. 60a

Fig. 60

and even pomp, if nicely finished.)

Whatever shape the platform for your treehouse assumes, it will have a perimeter. At the perimeter for unsupported spans of 7 feet or less, a 2x6 board is sufficient, unless you are planning to use a treehouse for uncommonly heavy activities. A 2x8 is strong enough for an unsupported 10-foot span, a 2x10 will safely span 14 feet. For floor joists, 2x4s are adequately strong, unless you have open spans of more than 8 feet, in which case use 2x6s.

If you decide you want walls and a roof, beyond what the tree naturally provides, put them up just as you would on any other frame structure, except that you might find it easier to complete the walls on the ground and haul them up, prefabricated, with a block and tackle. But be sure that the walls and roof are well clear of any limbs that might blow about and damage them.

You can, of course, use any roofing method or material you like, or that best suits your tree and treehouse. The common roof types are discussed in Chapter 3 on garden sheds.

If you don't want walls, you'll probably need a railing. See the directions given in the Playhouse chapter, page 128.

GREENHOUSES

Greenhouses

It is early April. A spell of warm weather has brought the daffodils and the forsythia almost to the point of bloom; the flowering trees are similarly poised, teetering on the brink of spring. It is all a little like a symphony orchestra tuning up for the opening chord.

The first notes are heard in my neighborhood at the very end of February, when the first green shoots appear. Then the willows begin yellowing in mid-March— a muted foretaste of things to come. Toward the end of March, buds begin to form on all sorts of things, and there are a few nice spells of warm weather here and there.

Finally everything is ready. The bows are poised over the strings, lips puckered to the horns, the conductor has his arms raised and head up, waiting for the signal from the Big Conductor in the Sky.

That's when it generally chills off and snows, and we have to wait another few weeks for the blossoms we want to see so badly.

But it doesn't have to be this way! Just as a good hi-fi can free you from the programs and schedules of the concert hall, so can a greenhouse provide you with spring, almost whenever you want it. If you want daffodils in November or tomatoes in January, you can have them.

Although many gardeners take full advantage of the capability of their greenhouses all year round, many others choose not to. It all depends on what it is you like to grow, and where you live; the further north, the longer the greenhouse season. People with glassed-in windowboxes and lean-to greenhouses—that is, greenhouses attached to the house—frequently use them only as winter gardens. As soon as the weather is mild enough in the spring, everything is put outside

Green·House·with·Aquarium·   by Maurice B. Adams, A.R.I.B.A.

Courtesy of Peter Reimuller—The Greenhouseman, Santa Cruz, CA

This geodesic dome greenhouse has fiberglass panels in a redwood frame. Notice that the door consists of a double triangle panel.

TYPES OF GREENHOUSES

A greenhouse can mean anything from a cold frame out in the garden to the Olympian extravaganzas shown on pages 161 and 163. In between are the greenhouses most people will be interested in. Smallest, but least expensive and easiest to install, are glassed-in window boxes, sometimes called *window greenhouses*, which may be bought as finished units from several of the firms listed at the end of the chapter. There are sizes for single and multiple windows, and generally they require no alteration of the house, although in some cases it may be desirable to remove the original window sashes.

The removal, of course, will leave an unfinished, raw hole in the wall, which must then be dealt with. If you wish, you can leave the face moldings of the window frame in place, and simply cover the tracks in which the window sashes moved with lattice of the required width, or 1/4-inch plywood cut to fit. Alternatively, the opening can be treated the same way you might treat an interior alcove or achway, finished with wood trim or molding, with plaster and

until fall. To be sure, you can't grow all orchids outside in Boston, say, even in summer, so you have to select your plants with some care if you plan on this kind of program.

Some gardeners only use their greenhouses as forcing beds in the spring, to get a head start on the growing season, and again in the fall, to protect from the early frosts sensitive plants that have not yet completed their cycle. (Used in this way, you may,

depending on where you live, be able to get along without a heater.) Others use the greenhouse's ability to control not only temperature and humidity, but hours of light as well, to grow vegetables out of season, or rare show plants like orchids or tropical succulents—all entirely within the greenhouse. In fact the growing requirements of any plant in the world can be met in a greenhouse.

PLAN

The wealthy English of the 19th century didn't do things by halves, as this flagrantly exuberant "Anglo-Japanese" greenhouse of the 1880s demonstrates. Its elegance is exceeded only by its staggering pomposity—but wouldn't it be fun to have?

lath, or plasterboard.

Some people leave the window in place, either to control temperature and humidity or, if they plan not to use the window greenhouse during the coldest part of the winter, the added layer of glass provides useful insulation. If the the window is double-hung, access to the whole interior of the greenhouse is still possible.

In that they are built on to the side of the house, *lean-to greenhouses* have a lot in common with window greenhouses, but they are much larger. They can be built on without altering the house, either without any direct access from the house or using a preexisting house door as entry to the greenhouse.

The first and most important consideration with a lean-to greenhouse is that it be placed on the side of the house where it will get the kind of light best for the plants you plan to grow. For specific information on this subject, have a chat with your local greenhouseman, nursery gardener, florist, or the dealer from whom you buy your greenhouse, if you decide not to construct your own. When you plan, remember that the sun describes a lower arc in the winter, and that trees that completely shade an area in summer may, once they lose their leaves, provide very little shade at all.

The least alteration to your house will be needed if you can erect the lean-to over an existing outside door. If you equip the lean-to with a door (most of the commercial ones can be ordered with or without an outside door), you can still easily get outdoors from inside the house. If you want the lean-to to provide a winter garden for a living area in your house, you will probably have to remove a wall section, or convert an existing window into a door. If effective temperature and humidity is to be controlled,

some kind of closing must be provided. Sliding glass doors can often be ideal. Most plants like both a lower temperature and a higher humidity than you would find comfortable in your living quarters.

Cutting a section out of an outside wall is not especially difficult, but you must know what you're doing. Outside walls are load-bearing walls, and adequate alternative support must be provided to compensate for the lost studs; this can be done in a variety of ways. Electric, heating, and water services must also be taken into account. If you are a novice at this kind of alteration, best get the help of a professional.

Finally there are *free-standing greenhouses* of all sizes, which can be put up wherever you choose in your garden. Many of these are no larger in area than an average lean-to greenhouse, but the fact that they are free-standing provides some advantages. For one thing, plants that like it can have sun all through the daylight hours. Others do best with just morning sun or just afternoon sun, and they can be catered for by selectively covering some of the greenhouse glass.

FRAME MATERIALS

Greenhouse frames are commonly made of one of three different materials: extruded aluminum, galvanized steel, or wood.

Aluminum is the best. It lasts virtually forever and requires no care, as it won't rust or rot. But it's expensive.

Galvanized steel doesn't require much care, but it will eventually rust. Still, if the zinc coating is reasonably thick, you can expect a good twenty years from it, which may be as close to forever as you are interested in.

Wood is still favored by some. I suspect the main reason is that people just *like* wood better than aluminum or steel. It will eventually rot, of course, although there are wood-framed greenhouses standing here and there that were built in the last century. It needs painting or varnishing from time to time, and painting a greenhouse frame is undeniably tedious.

GLAZING MATERIALS

There are also three materials now commonly used for the glazing. Glass, of course, is one. The others are fiberglass and plastic, the latter transparent or semi-transparent, rigid or flexible. Each material has its advantages and disadvantages. Glass is the only one of the three that time doesn't affect at all. It is also the most transparent and relatively easy to repair when it breaks. And break it can; hailstorms can be a disaster. What's more, there is something about a glass greenhouse that makes the fingers of your average small boy itch for a stone—I hope Mr. Welzenstein down the street from where I grew up doesn't have a long and vindictive memory. Glass is also relatively expensive, particularly if you choose double-strength glass to minimize breakage.

Rigid plastic comes in several thicknesses. The thicker the better for a greenhouse, generally, although even within a thickness class there are quality differences. The best quality plastic (about as expensive as glass, by the way) is quite transparent, yet much harder to break than glass. Unfortunately it isn't very long before the glasslike transparency goes, little by little, until everything seen through it looks blurred and overexposed. To be really clear, a transparent material must be perfectly

Courtesy of Lord & Burnham

A glassed-in window box has enough room in it to satisfy the needs of many gardeners.

smooth on both sides. Small irregularities in either surface will disperse the light and cause blurring. Well, this plastic, being softer than glass, is gradually eroded on the outside by dust in the wind, grit in the rain, and fogged by ultraviolet light from the sun.

At the other end of the plastic scale there is the thin flexible stuff that hardware stores sell people to make jerry-built storm windows. If you want to recover your greenhouse every year (some professional greenhouse gardeners do, by the way, and claim it's the cheapest way to go), you can get by with it quite satisfactorily.

In between there are several grades and thicknesses of plastic sheet available which, if properly applied to the frame, can last on a greenhouse for several years, and it's still lots cheaper than glass or really top-grade plastic. To be sure, all flexible sheeting is only semitransparent to start out with, and the sun's ultraviolet rays and atmospheric grunge will eventually discolor it and make it brittle. Strong winds, which can cause destructive flapping, twigs, and plant poles incautiously lifted inside can also cause problems with plastic sheeting. It won't break, but it's certainly easy to poke holes in it. The lightest grade you can reasonably use on a greenhouse is 5 mil, and it will likely require annual replacement. Ten mil plastic should last for two or three years, if it is put on carefully, using lath strips as battens. Ten mil costs less than twice as much as 5 mil, and so could be cheaper in the long run, particularly if you reckon in labor costs.

Fiberglass is rigid, stronger even than glass, is virtually unbreakable, and is adversely affected only very slowly by ultraviolet radiation, if at all. It is, however, not transparent, only translucent. Some green-housemen say translucent is better, that diffused light is better for plants than direct sunlight. Certainly lots of fine and healthy plants have been grown in fiberglass greenhouses. On the other hand, you may like to see out of a greenhouse when

---

HYDROPONIC GARDENING

A greenhouse hobby that has become popular in recent years and might interest you is hydroponic gardening. It means growing plants in a sterile medium (small marble chips are often used), with all nutrients added in the water. In this way it is easy to select just the density of growing medium and kind of nutrition best for each plant. If you are new to greenhouse gardening, it's possibly something you ought to look into before you select the type of greenhouse and kinds of benches you want. The requirements of hydroponics are a little different from those of dirt gardening, and some planning ahead should be a good idea. You can get information on the subject from the Plantworks (address at the end of this chapter), or from the following recent books:

Dickerman, A. C., *Discovering Hydroponic Gardening,* Santa Barbara, Woodbridge Press, 1975
Loewer, H. P., *The Indoor Water Gardener's How-to Handbook,* New York, Walker Press, 1973.
Sullivan, G., *Understanding Hydroponics,* New York, Sullivan, 1976.

---

you're inside, and to be able to look in when you're outside.

Before you build your own greenhouse, or buy one in a kit from one of the suppliers listed on page 180, there are a few other variables to which you must give consideration.

LIGHT CONTROL

The function of a greenhouse is to exclude the weather in all its forms, yet to capture the radiant energy of the sun, both to generate ambient warmth in the greenhouse, and to make possible the photosynthesis by which plants live. Often in summer more energy is captured in the form of heat than is desired, so any greenhouse must have a means for overheated air to escape. In a lean-to or free-standing unit, that is usually provided by opening panels in the roof.

In addition, it will probably be necessary on hot summer days to arrange some kind of shading. Greenhousemen frequently find it satisfactory and sufficient to whitewash the roof panes, although most firms that sell greenhouses also sell adjustable blinds that fit against the glazing on the inside. You can make perfectly adequate blinds using ordinary roller window blinds,

A lean-to greenhouse on the side of a home can harmonize nicely, and provide plenty of space for amateur horticulture. This one has a concrete-block lower sidewall, and rattan porch blinds to control the light.

Courtesy of Lord & Burnham

mounted near the roof peak, and controlled by cords run through a screw eye at the top of the wall. Best for this purpose is the translucent, mildew-resistant material sold in drygoods or curtain stores for use in bathrooms.

It is not necessary, by the way, to adjust your blinds frequently. Because the sun moves across the sky, it is enough, usually, to pull every other blind, thus limiting the exposure of each plant to direct sun's rays.

The placement of your greenhouse will to some degree control the kinds of plants you can grow. A spot that has direct sun most of the day is ideal; you can shade it to suit your particular needs. This kind of site selection is easier for a free-standing unit than for a lean-to, which must be placed with some consideration for the house against which it is built.

If you plan to grow plants outside their normal season and need to extend the period of daylight artificially, you can buy special fluorescent lights that will fool your plants into believing it's a summer day when it isn't.

CLIMATE CONTROL

Unless you live where frosts are rare, you will also need a heater. Good heaters—electric, gas or oil—are available through greenhouse-supply companies; some are listed at the end of the chapter. Greenhouses aren't as hard to heat in winter as you might think, because the sun does a lot of the work for you. Still, it isn't always sunny, and even when it is, the sun needs help in any weather much colder than chilly.

Your dealer will help you decide the size and type of heater you need, depending on your particular conditions. There are so many variables of size and type of greenhouse, exposure, weather, insulative value of the walls, and plant-type preferences, not to mention ways to heat a greenhouse, that any recommendations I could make would be meaningless. Anyone interested in a greenhouse simply has to tailor it to his needs and climate, and there is no shortage of local commercial talent to help with that. But I can outline some of

You can buy greenhouses, assembled or in kits, which can do most greenhouse jobs very inexpensively.

the heating options.

For a lean-to, consider using a heat outlet from your house furnace; it must have its own controls, however, since your plants like a different climate than you and your family do.

For a free-standing greenhouse, the easiest heating arrangement usually involves an oil-fired unit fed from an elevated oil drum or fuel tank immediately outside the walls. Coal or wood-burning heaters are cheaper to operate, but more difficult to control—particularly to control automatically.

Outfitting the greenhouse will cost least if you choose a manually controlled heater and roof vents that you open by hand. Some additional expense (and well worth it, I feel) will get you thermostatic control of your heat. Yet more money will buy you electric motor arrangements that will open and close your roof panels at the touch of a button. For these, of course, you must have an electric supply. Then, if you like, the electric motors can be wired through another thermostat, so they open and close the vents when preset temperatures are reached.

WATER SUPPLY

Don't forget to lay in a frost-protected water supply. A greenhouse without running water is about as useful as a stove without saucepans. Easiest will probably be to run a pipe from the house water supply. Be sure to bury it below the frost line, if it is to serve a free-standing greenhouse. For this black plastic pipe is the best, if it is permitted by your local plumbing regulations. It never rusts or rots, and is inexpensive to buy and to install.

If you plan a walk-in greenhouse, whether lean-to or free-standing, two water outlets, possibly three, are useful. First you must have a hose connection for watering and spraying and hosing off the floor. Second, it is nice, if not essential, to have a spigot under which you can rinse out pots or wash your hands. The possible third connection is for an automatic humidifier, an aid which will greatly reduce the amount of time you have to spend watering and spraying.

FOUNDATION AND FLOOR

Plan your foundation well. A greenhouse that sags at one corner is going to mean a lot of broken glass, sprung fiberglass, or ripped plastic. Good, solid footings are in order. If your ground is soft, that may mean a whole perimeter foundation. The dimensions of the footings or foundation will, of course, vary with the size and style of the greenhouse that is to stand on

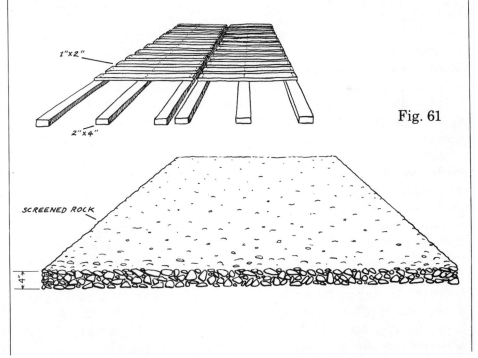

1"x2"

2"x4"

SCREENED ROCK

4"

Fig. 61

them. If you buy a greenhouse kit from a manufacturer, directions for the required foundation will come with it.

Generally a greenhouse does not require a permanent floor. About 4 inches of 1½-inch screened rock on top of the earth will do nicely. Lay 1x10 or 1x12 boards down to walk on, or build slatted walkways where you need them. Lay parallel 2x4s 8 inches apart, then nail 1x2s crosswise to them, spaced ¾ inch apart, as shown in Fig. 61, page 169. Make the walkways in sections— 4 feet long is usually handy, but tailor them to the size of your green-house— so they can easily be picked up and taken out when the underfloor needs cleaning. Wooden walkways are also good on a concrete floor.

For a lean-to greenhouse that is intended to form a winter garden off a living area, something more substantial may be desired. A reinforced concrete slab will serve nicely (see chapter on garden sheds, page 71), but don't lay it all the way to the foundation footings along the rear wall, *away* from the house. Leave a strip about a foot wide between the slab and the foundation, dig it out to a depth of about a foot, and fill it with screened rock, for drainage. Tilt the slab slightly towards the drainage strip when you pour and finish it. No other drainage should be necessary. Even large puddles on the floor will soon evaporate and will in fact help keep the air humid.

BENCH DESIGN

Give some thought to the benches—the tables on which you put the plants. You want them to be higher than a standard desk or table, so that you can work standing up, yet not so high that you cannot look down into a pot. Most people are comfortable with a bench about waist high. On the other hand, you may wish to sacrifice working comfort for more headroom for tall plants.

An ideal work surface for the benches consists of 1x2s laid on edge next to one another, but with about ¼ inch of space between them. This will provide a solid and stable surface, but one through which excess water can drip, and which can be cleaned off with a hose. Painting isn't necessary. With all four sides exposed to the air, the wood will dry out quickly and won't rot for years and years if ever. If you use redwood or cedar lumber, it will be virtually permanent. Fig. 62 on page 171 shows a simple bench design. If you have a table or radial-arm saw, and you find 1x2 hard to get in cedar or redwood, or if you want to save a little money, you can rip 2x4s into quarters. But remember, ripping (cutting a board lengthwise) requires a sharp blade if the cut is to be true.

Here is a way to build a good, simple bench.

1. Make a rectangle of 2x4s the size of the bench surface you want to have. For the sake of flexibility in the use of your greenhouse, make each bench between 3 and 4 feet long. Short benches are easier to rearrange or remove—should you need to make room, for example, for a very tall plant. A good working width is 18 inches, if it suits the interior of your greenhouse. Use 3½-inch common nails, three for each joint.

2. Decide how high you want the benches to be. For each bench unit cut four 2x4 legs that are 1-½ inches shorter than the working height you decide upon.

3. Tack-nail or clamp the legs in position inside the corners of the rectangle, the tops of the legs flush with the top of the rectangle boards.

4. Drill three 3/16-inch holes through the rectangle side boards into each leg: two holes an

inch from the lower edge of the side board into the wide side of the leg, and one hole an inch from the upper edge of the other side board into the narrow side of the leg. Number each leg, writing the number not only on the leg but on the rectangular frame where the leg is to go. Remove the legs and set them aside.

5. Cut as many pieces of 1x2 as your benches are inches long, each piece as long as your benches are wide. For example, if you have two benches, each 40 inches long and 18 inches wide, cut 80 pieces of 1x2, each one 18 inches long. Be sure you cut the pieces the *full* outside width of the bench top, *not* the inside width between side members.

6. Mark the tops of the long side members of the bench tops at inch intervals. Nail the 1x2 pieces onto the tops of the bench top side members, spacing them one an inch, lying on their narrow side. That will leave you a ¼ inch space between strips. Use 2½-inch finishing nails for this job. If you find that the strips tend to split, then drill the 1x2 pieces for the nails—but don't drill the 2x4!

7. Finish off the top by running a strip of 1x2 along the butt ends of the cross strips. Nail

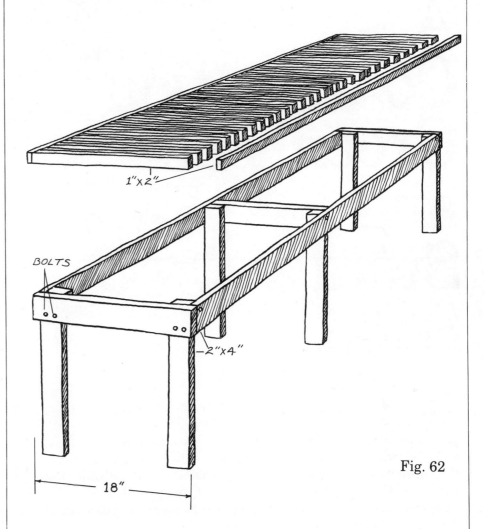

Fig. 62

it in place with 1¼-inch finishing nails.

8. Turn the tops upside down, and put the legs in place with ⅜-inch lag screws. Use a 4½-inch screw into the narrow side of each leg, and 3-inch screws into the wide sides. Put a flat washer on each screw.

A Combination Greenhouse and Storage Shed

A greenhouse doesn't have to be just for plants. Called conservatories or winter gardens, and attached to the house, they have frequently been favorite places to spend time on cold but sunny winter days. Alternatively, a detached greenhouse, with not much sacrifice of its basic function, can be arranged to serve both as garden shed and potting shed—but it does take some arranging.

You could, of course, haul some potting soil and pots into your greenhouse (*voilà*, a potting shed!) and put a rake and a spade on the door (hoopla, a garden shed, too!), but that's not a very satisfactory road to versatility; most home greenhouses just don't have the space to double as potting and garden storage sheds. The potting shed needs shelves for pots and for bags of potting soil and fertilizers and such, and ample drawer space for storage of seeds and bulbs and gardening gloves and all sorts of useful carry-on, see Fig. 63. It's difficult to arrange all that in a small greenhouse; shelves are difficult in any greenhouse, as the glass walls just aren't suited to hanging them.

The garden shed needs plenty of space for the various machines that gardeners use today, along with more floor and wall space for a wheelbarrow, spreader, hose, rakes, spades, hoes, and so forth. A place to hang a ladder is nice, too.

All this would seem to say that if you want a greenhouse, build a greenhouse, if you want a potting shed, build a potting shed. Not so. You simply need to plan carefully. You can have greenhouse, garden shed, and potting shed all in one, if you plan it in detail and make maximum use of your space. With a little careful designing, a combination shed and greenhouse can be arranged to suit your needs nicely.

Here is a somewhat abbreviated description of a combination building I built some years ago, mainly of used materials. It is abbreviated because the framing was mostly conventional, and so dealt with elsewhere in this book, and because my particular needs then are bound to be different

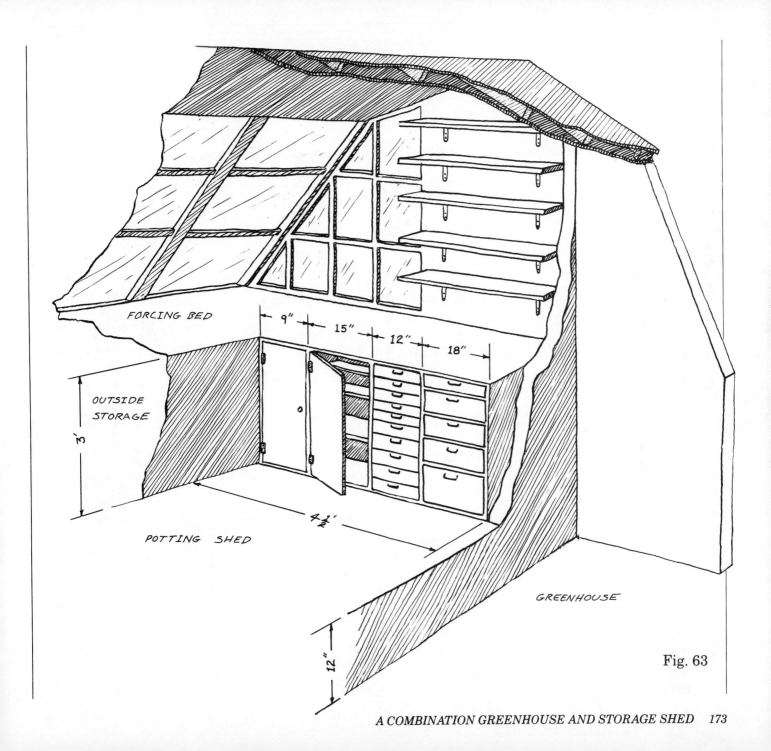

FORCING BED

OUTSIDE STORAGE

3'

9" 15" 12" 18"

POTTING SHED

4½'

12"

GREENHOUSE

Fig. 63

A COMBINATION GREENHOUSE AND STORAGE SHED   173

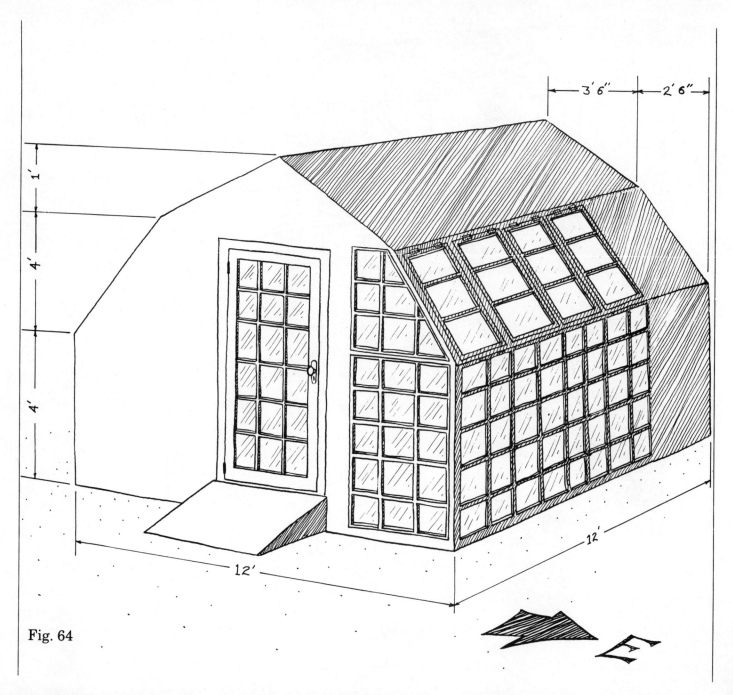

Fig. 64

from yours now. What I think may be of interest about the building are the way it was planned, its layout and organization, and a few special framing techniques, including one or two forced on me by the odds and ends of lumber, storm windows, and other used materials I had to work with.

Before I touched wood or hammer I drew the outside walls, the interior divider wall, and the floor area on graph paper. Because I wanted to be able to sheathe the outside with minimum wastage from 4x8-foot plywood sheets, I decided on 12x12 feet for exterior dimensions, and allowed for wall thicknesses in planning the interior. I decided on a gambrel roof because it would permit headroom of at least 7 feet throughout the working area, with an overall building height of only 9 feet. Low buildings are easier to build (see Fig. 64).

Then I measured the various tools and machines I wanted to be able to store, and drew them in, hanging or standing where I would want them to be. From this exercise I learned there wouldn't be room to hang my ladder along the interior wall, so I decided to make a rack for it under the floor on the forcing-bed side. On the end wall, under a window, I laid out the dimensions of the cabinet I wanted to build in, and decided on the number and depths of the shelves.

In laying out the floor I found I had not nearly enough room to store the garden machines, such as mower, tiller and snowblower, too heavy or bulky to hang up. Rather than build a larger building, I looked for alternative space and found it, under the forcing bed. I decided to close that space off from the inside and provide access through individual outside doors. With the rest of the machinery thus taken care of, there was just room on the floor inside for the garden tractor. It would be a little awkward to work in there with the tractor, but that seemed less troublesome than enlarging the building.

Then, to be sure I had done all my arranging and planning properly, I built a model—1 inch to the foot—of the building, using shirt cardboard, white glue, and transparent tape. I was glad I did. First, it was fun to make the model; finishing it gave me almost as much satisfaction as subsequently finishing the building. Secondly, I discovered that the door I had planned between the forcing bed and greenhouse sides swung open just nicely to smash a pane in the greenhouse glass. Hard to foresee on paper, but easy with the model. When I thought about it, I couldn't see why I needed a shutting door between the two sides of the building to begin with, so in the final execution I left it off.

In the greenhouse section the glass extends right down to the ground, so the area under the bench may also be used for plants. The rear quarter of the greenhouse is without glazing, to provide a shaded area for the initial growth of bulb plants, which need that kind of environment. That area can also, of course, be used for potting.

I decided against a made floor in the greenhouse, and settled instead on 3 inches of crushed rock on the bare earth, and a walkway made of three 1x4s, with ½-inch gaps between them, nailed at 3-foot intervals to 1x4 cross pieces. The idea I had had to store my ladder under the floor on the workroom side looked doomed for a while, as there wasn't room for it and floor joists—even just 2x4s. In the end, after much agonizing, I plunked for 1-inch plywood for that floor. It's expensive, but stiff enough that I figured I could get along

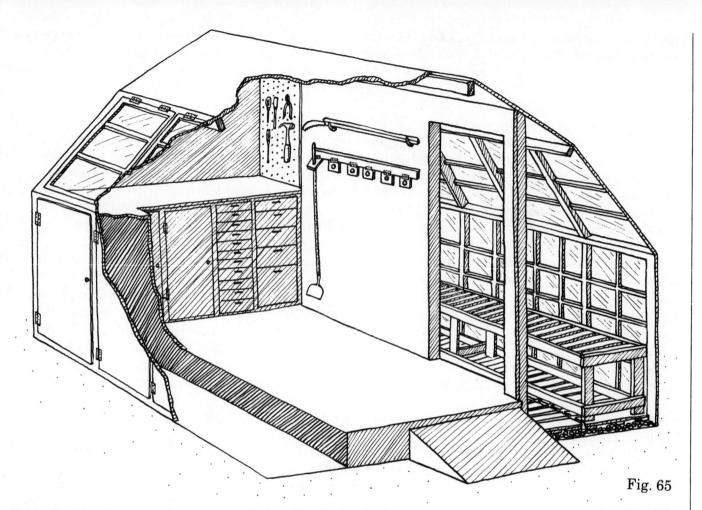

Fig. 65

without joists. Often in building if you need to weaken somewhere, you can compensate for it by stiffening somewhere else.

The other side of the building has a triple function. The high 3x12-foot counter under the sloping glass of the lower roof section is a forcing bed for seedling plants in the spring. The 36 square feet of counter area, which get the full afternoon sun, is plenty to supply the needs of quite a large flower or vegetable garden. At the rear is a workbench containing a number of shallow drawers for bulbs and seeds, and a few larger drawers for the inevitable odds and ends. There is also a cupboard with 3 shelves in it, and a cabinet for a garden sprayer. Over the bench, to the right of the window, are five shelves for pots, soil, fertilizer, and so forth. The space could, of course, have been divided up in any number of other ways.

The long dividing wall isn't really necessary at all, except to provide space to hang up all manner of garden tools and hand tools, and to keep one from stumbling down the step into the greenhouse side (see Fig. 65). Without the wall that height differential in the floor would be hazardous. Still if the ladder is to be stored under the floor, and to allow useful height in the outside storage compartments and at the same time keep an acceptable working height at the forcing bed, the height differential is unavoidable—unless work and plant headroom were to be sacrificed to raise the floor on the other side. Notice especially the no-slip hangers for all tools with round wooden handles, Fig. 66. If you make hangers like them, be sure to set them high enough so you can slide the tool out the bottom.

Four small doors in the outside wall, underneath the forcing bed, give access to compartments for larger items such as lawn mower, wheelbarrow, spreader, and the like (see Fig. 67, page 178). The rearmost of those four doors opens onto a fuel-storage locker, which has a ventilated floor and walls lined with plastic sheeting, to keep fuel fumes out of the rest of the building.

At the rear of the building,

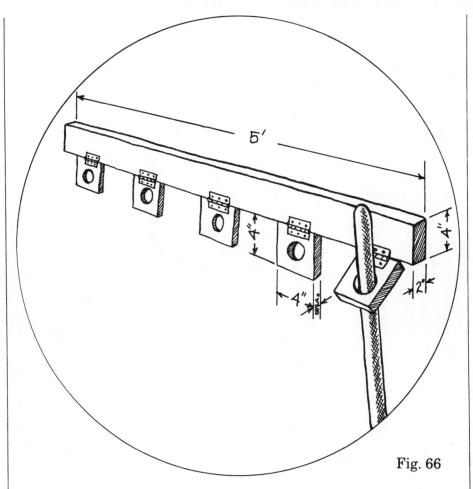

Fig. 66

at ground level under the north window, is a small door for the ladder; a 12-foot stepladder or a 20-foot extension ladder can be accommodated. The roller rack screwed to the underneath of the 1-inch plywood floor is a standard item that can be bought from a hardware store.

In the building as shown all the angled glass panels are hinged at the top, so they can be propped open for relatively rain-free ventilation. They were made from discarded wooden storm windows, as were the glass base panels on the long, east wall of the greenhouse (see Fig. 68).

VENTILATED
FUEL STORAGE

Fig. 67

The rear window, all the siding and framing lumber, and the front door were bought from a used-lumber yard, very inexpensively. Some glass had to be replaced, of course, and it was necessary to use new materials for some of the building—notably the potting bench, the front, south glass wall of the greenhouse, the exterior compartment doors, and the roofing—but the total cost of the building was just a few hundred dollars. That is no more than the cost of a standard steel storage building of similar size, with a similar foundation, and very much more useful, attractive, and almost certainly longer lasting.

Given different needs and a differing stock of available materials, the building could

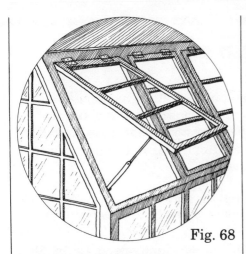

Fig. 68

have turned out in lots of different ways. If a more conventional greenhouse had been the goal, more wall and roof area could have been glassed and the external storage compartments could have been eliminated—thereby gaining that much more interior room into the bargain.

Going the other way, the greenhouse area could have been done in solid walls and used for additional storage space or, eliminating the interior wall, to enlarge the working area; practical if the floor were leveled. Above all, one thing should be clear: You can plan more satisfactorily and build to correspond more nearly to your needs if you design carefully and build your own building than if you buy a ready-built structure,

whether a shed or a greenhouse—and it needn't cost you much if any more.

The one feature of the frame of this building that isn't described elsewhere in this book is the gambrel roof. A gambrel roof is an attractive option for a small building. It looks nice, and it provides maximum headroom for minimum building height within the constraints of needed roof pitch.

If you want to put a gambrel roof on any small building you put up, start the above-the-floor

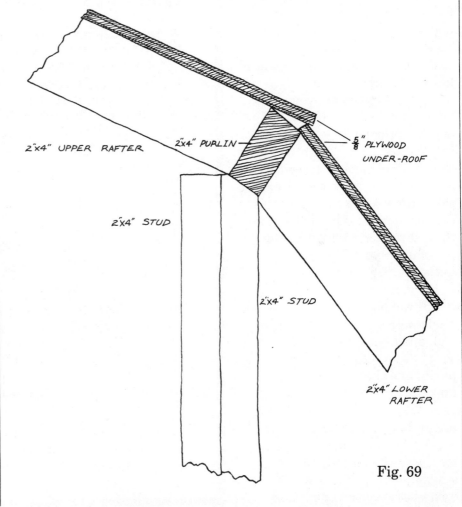

Fig. 69

framing with the end walls. You can either build them conventionally, and frame for the gambrel only in building the rafter, or you can trim the tops of the studs to conform to the shape of the roof—in effect eliminate the wall plate, letting its function be assumed by the end rafters. The latter option is usually preferable, as it will make the placement of windows or doors in the end walls easier if low eaves are planned.

At the roof peak, you can either use gussets to hold opposing rafters together or nail them to a ridge pole; similarly, at the gambrel you can use either gussets or a purlin, which is what the equivalent to a ridge pole is called when it is found elsewhere than at the topmost ridge. Not uncommonly builders use gussets at the peak and a purlin at the gambrel.

If you use purlins, double the studs supporting them at either end, and trim the doubled studs so that they support all the purlin and the very end of the upper rafter, see Fig. 69 on page 179. Intermediate rafters are arranged in the same way, except they need not be beveled for the stud, and they can be staggered one rafter's width apart between the upper and

GREENHOUSES AND GREENHOUSE SUPPLIES

This list of companies that sell prefabricated greenhouses and greenhouse supplies is not by any means exhaustive. But the companies listed offer variety in types of greenhouses, materials, and cost. Prices range anywhere from $200, depending on size, material, and "extras," to thousands.

W. Atlee Burpee Co., Warminster, Pa. 18974
Burpee makes a free-standing aluminum and glass greenhouse, 8x12 feet and a lean-to model, 6x12 feet.

Environmental Dynamics, P.O. Box 996, Sunnymead, Ca. 92388
A variety of greenhouses, mostly with steel frames. The smallest is 4x5 feet available with sheet plastic or fiberglass glazing. The largest (steel and fiberglass) is 8x24 feet.

Everlite Aluminum Greenhouses, 14605 Lorain Ave., Cleveland, Ohio 44111
Everlite offers a selection of glass and aluminum greenhouses, lean-to and free-standing. The smallest free-standing model is 7½x8½ feet. They go up from there to a width of 17 feet, and as long as you want, in 10-foot sections.

Gothic Arch Greenhouses, P.O. Box 1564, Mobile, Al. 36601
Gothic offers several sizes of free-standing greenhouses, from 12x12 feet to 22x40 feet. Four-foot sections can be added to the length to achieve further variation. Materials are redwood and fiberglass.

Janco Greenhouses, 10788 Tucker St., Beltsville, Md. 20705
Janco offers a lot of different styles of glass and aluminum greenhouses, with free-standing models ranging in size from about 9x11 feet up to 21x52 feet. Extra sections can be added to increase the length of all the models.

Lord and Burnham Greenhouses, Irvington, N.Y. 10533
Burnham makes greenhouses out of glass and aluminum, and also has sizes, free-standing, from small (8 feet 6 inches x 9 feet 9 inches) to very large (21 feet x 51 feet 6 inches), and there are lots and lots of equipment and style options. In addition, Lord and Burnham will make any kind of aluminum-framed glass construction you want, on special order.

McGregor Greenhouses, 1195 Thompson Ave., Santa Cruz, Ca. 95063
McGregor makes theirs out of redwood and fiberglass. These appear to be basic, no-frills greenhouses. Free-standing models range from 4 feet 2 inches long to 16 feet 5 inches long, all by 6 feet 10 inches wide. Extension sections are available.

The Plantworks, 100 Progress Parkway, Maryland Heights, Mo. 63043
The Plantworks offers two styles of steel-and-fiberglass free-standing greenhouses. One has a rounded top like a covered wagon, the other looks a little like a miniature edition of the chapel at the Air Force Academy. It's attractive. Plantworks is particularly enthusiastic aobut hydroponic gardening, and their greenhouses are designed with this kind of gardening in mind. The chapel-shaped model is 10 feet 9 inches wide and 14 feet long, and extensions are available.

Peter Reimuller Greenhouses, 980 17th Ave., Santa Cruz, Ca. 95062
Reimuller has redwood frames and glazing either of polyethylene or reinforced fiberglass. Reimuller's greenhouses come in more interesting shapes than most, see the photo on page 162. There are also more conventional greenhouses in sizes up to 16x8 feet.

Sturdi-built Manufacturing Co., 11304 S.W. Boones Ferry Rd., Portland, Or. 97219
Studi-built offers quite a wide variety of shapes and styles. They have redwood frames and either fiberglass or glass glazing. One of their greenhouses, an 8x10-foot "Sun-Bon," is glazed with large sheets of milky, translucent fiberglass that looks opaque from the outside. It is built in the style of a Japanese tea house. Another is round, (actually 24-sided) with outward-sloping walls and a shallow conical roof, and is available in 10-, 12,- and 15-foot diameters. Other more conventional models are available too.

Turner Greenhouses, Highway 117 South, Goldsboro, N.C. 27530
Turner offers basic, conventional, steel-frame greenhouses in sizes from 8x10 feet to 14x46 feet, glazed either with plastic sheeting or fiberglass.

Vegetable Factory, Inc., 100 Court St., Copiague, N.Y. 11726
The Factory features a double-layer thermal glazing of reinforced fiberglass bonded to an aluminum frame. The design is said to substantially reduce the energy required to heat in winter, while the light-diffusing qualities of the two layers of plastic are supposed to eliminate the need to shade in summer. Shapes are conventional. The design is said to be especially effective for growing vegetables.

lower roof slopes, so that it isn't necessary to toenail them to the purlin. The end rafters, which can't be staggered, you will have to toenail.

When you put on the underroof, be sure the covering of the upper slope slightly overlaps the lower slope, as shown.

GAZEBOS

What Is a Gazebo?

A well-designed and well-placed gazebo, or garden house, is an addition to almost any backyard or garden. It's nicest, of course, if it can be a focal point in a grand sweep of greensward and carefully casual planting, or perhaps set off by a little waterfall and some topiary plantings. But even if all you have is an ordinary suburban backyard with a tree or two, and some grass that you have struggled to make attractive, a gazebo, if you're reasonably careful about its design and placement, will look just dandy.

There is just one terminological matter which needs attention here and now. Depending on pretensions and dialect area, little garden shelters are commonly known as summer houses, screen houses, belvederes, pavilions, bowers, arbors, follies, garden houses, or gazebos. Dictionaries aren't a lot of help in making clear distinctions between these various terms, although most people seem to feel that each one means something slightly different. Maybe so.

To me a *summer house* is a house in which people live in the summer, in the woods or by the shore. A *screen house* is a gazebo with screened sides or windows. These are distinguished from *belvederes* and *pavilions*, which have no sides at all, but usually only a roof supported on columns over a raised platform. Belvederes and pavilions are usually a lot more pompous and showy than garden houses, and perhaps for that reason those

Courtesy of Koppers (kits available)

Notice particularly the floor and underroof treatment of this redwood gazebo. But unless you can buy the rafters ready-curved, I doubt that they'd be worth the work they'd take to form.

names are sometimes given to garden houses, by way of wishful thinking. A *bower* or *arbor* is a wooden or metal framework over which plants have been trained to grow. A mature bower can be difficult to distinguish from a vine-covered gazebo, but in winter, when the leaves are off, you'd see the difference right away. A *folly* is an artificial ruin, a structure designed to look like the romantic bare remains of a destroyed castle or decayed abbey. Follies are best viewed in the moonlight, in the company of someone with whom you are passionately in love. In other circumstances, they have a tendency to look absurd. *Garden house* is a nice, neutral phrase, and not subject to much confusion, although one source has told me that it's a term sometimes applied by over-decorous people, particularly in the American South, to their privies. In this book it means the same thing as gazebo.

Gazebo itself is an interesting word. A lot of people avoid calling

As this gazebo demonstrates, remarkable things can be done with ornamental wrought iron. Notice the central spiral staircase.

anything a gazebo because they're afraid they might say it incorrectly. If they would be inclined to say "*gaze*-bo", their worries are well-founded. That's wrong, and no question. Dictionaries list both "guh-*zee*-bo" and "guh-*zay*-bo", although the former is generally listed first. Perhaps that's because "guh-*zay*-bo" is also a rather dated slang word for "guy."

The origins of the word are obscure. Some have suggested that it is an intentionally comic formation of English *gaze* and Latin first-person verb ending, -*bo*, creating a sort of Anglo-Latin word meaning "I shall gaze." That granddaddy of all English dictionaries, the Oxford English Dictionary, however, suggests that *gazebo* is a corruption of an Oriental word, as are *tea* and *chop suey*.

It seems that garden houses have been around almost as long as houses. Almost five thousand years ago, they were a regular feature of the gardens of ancient Babylon and Egypt, and they were popular in China of the Shang Dynasty, about 1400 B.C. Pompeii had its garden houses, as did Rome.

After the decline of the Roman Empire, garden houses seem to have declined in vogue.

This 18th century German garden house, complete with formal columns and a thatched roof, is nicely situated, away from the main house, at the edge of the woods. If you like the idea of a porticoed entrance, used lumber yards are a good source for columns.

There is an exception to this, however, among the clergy. While most people were struggling to survive in post-Roman Europe, many brethren had, if little else, time to sit and contemplate, and not surprisingly, garden shelters of various kinds were common features of monastery gardens.

Often, of course, they functioned as shrines where monks could pray to particular saints, but occasionally they seem to have existed purely as decorative places for solitude.

With the Italian Renaissance, the ornamental garden shelter came back into style, although it

This gazebo is built almost entirely of lattice strips. It would not be hard to duplicate, and it is undeniably attractive.

had caught on in England. No proper Englishman's garden was complete without a garden structure of some kind, and quite a few were extravagant to the point of scandal.

It wasn't until well into the eighteenth century that the Germans got involved in garden shelters, but when they did, it was in a big way. First came a rash of classical temples, but with the rise of romanticism, from about 1790, there was a rush to build the most bucolic shelters possible—some of them outdoing in rusticity anything a peasant would willingly have called home.

Because the late eighteenth and early nineteenth centuries were still a period, by and large, of the very rich and the very poor, it isn't surprising that there were relatively few modest garden shelters; it seems to have been either extravaganza or nothing. The less pompous garden shelter, the gazebo, as most people think of it today, began to become genuinely popular in the second half of the nineteenth century, and has continued so, with the exception in America of two dips in popularity, from the second of which we are just emerging.

The detached garden house enjoyed a striking boom in the

seems to have been a while before the idea, along with other features of the newly found bourgeoise good life, made its way over the Alps into northern Europe. By the early part of the seventeenth century, however, fanciful pavilions and mock temples were beginning to spring up in France, and the whole idea

United States in the period from 1850 to 1900. This is easy to understand, for these years correspond with the rise of a large, relatively prosperous class, whose members were anxious to parade their new wealth. A gazebo, along with such other items of conspicuous consumption as a grand house and a victoria with a matched set of bays, was simply one of the signals to the rest of the world that the person had made it. That the gazebo subsequently turned out to be a pleasant place to sit and enjoy solitude and one's own thoughts after a busy day was for many newly rich gazebo builders an unexpected bonus.

Then, around 1900, people stopped building gazebos. The reason, at least in part, lay in the rapidly waxing popularity of the attached porch. What is the use, people probably wondered, of a little, cramped structure off in the garden when it was possible to sit on a spacious porch, perhaps even protected from insects by the newly available metal screening? Few middle-class homes built in America before 1870—save in the South—had porches; but by 1900 the tables were turned, and few built after that date didn't have them.

Much was lost in the change of style. The porch couldn't offer the sense of intimate privacy that a garden house could, nor did it do anything to enliven the view of the garden, but in the eternal quest for modernity, such amenities were overlooked. The gazebo, however, soon enjoyed a renaissance. By the 1930s, even people whose houses included fine screened-in porches were beginning to build garden houses once again.

After World War II, gazebos suffered another lapse in popularity, this time to be supplanted by the patio. Of all the styles in home amenities over the years, I find the patio one of

Courtesy of Koppers (plans available)

This redwood garden shelter can be moved around, which could be an advantage in some backyards. It would make a good pool-side shelter.

the hardest to understand. As far as I can see, the only advantage a patio has over a plain grass yard is that your chair legs don't sink in.

For one important thing, a patio is worthless when it's raining. Anyone who has never sat on a porch and watched—and smelled—a heavy summer rain cascading down, and perhaps been chilled every now and then with a gust of spray, has missed an important and extremely pleasant experience. Well, it's even better in a gazebo. You're all but out in the middle of it, only you're not getting wet. The contemplation of just one good summer storm, alone or in the company of just one person you like a lot, and perhaps with a good warming drink, will abundantly repay the costs of a garden house. There is no substitute.

The fact is, a gazebo offers a lot of attractions of a timeless sort. There is, for one thing, the peculiar satisfaction in something that is utterly impractical, that exists solely for the pleasure it can give. In this sense, a gazebo is like a work of art.

Another of the gazebo's more appealing attractions is that it is small. Smallness is something that for one reason or another fascinates people. This book is about *little* houses, and the garden house is a pure example. The big lived-in house, full of conveniences as it may be, has to be shared with other members of the family. Your own room may be private enough, but it remains undeniably a part of the shared house.

A gazebo is different. When you are in it alone, it is yours exclusively, outside wall to outside wall, roof to ground. Furthermore, because you can see every corner of it from any place inside it, you *know* it's yours alone. That's something you can never be entirely certain of in a multiroom house, whether indoors or on a porch.

CHOOSING A LOCATION

Where you put your gazebo will, of course, considerably affect its value as a garden ornament, or as a place to sit and enjoy the surroundings. Basically, there are two ways to approach the problem. Either you can make the gazebo the focal point of your entire garden or you can blend it into the natural features. You may be able to do both.

If you have a lawn with little in the way of trees and shrubbery, you don't have much choice. You can't blend a gazebo in with the lawn, even if you have a lot of dandelions. If you can do it, put it at the top of a little rise—you might even consider *making* a little rise with fill and topsoil to put the structure on. Or you might place the garden house right at the back of the yard, and lay a flagstone path from your back door directly to it. In any case, keep the gazebo well away from your garage or toolshed, or the like. The styles probably won't clash, but what will clash is the juxtaposition of the absolutely practical with the absolutely impractical.

If, on the other hand, you have a back garden with lots of greenery, plenty of shaded areas, then the second approach is probably best. Blend it in. Nestle the garden house in between a couple of lilac bushes, for example, or let it just protrude above a big, round forsythia bush. Don't put in right next to a tree, particularly if you pick a generally vertical style of gazebo, as the tree will draw too much eye interest, and the gazebo will look as though it had just been put down there temporarily.

SCROLLWORK ORNAMENTATION

The heyday of the gazebo in America occurred between 1860 and 1900. Not so incidentally, this was also the high point in American architectural ornamentation. As a consequence, some of the nicest gazebos are highly ornamented with scrollwork and finials.

The most usual place for scrollwork is under the eaves, on the gable ends of the building. But it may also be used to cap windows and doors, or outline the frame of the structure. How much you use is entirely a matter of taste and personal preference.

Sometimes scrollwork was highly regular, perhaps nothing more than a series of pendant circles or scallops, other times you find highly intricate designs, incorporating flowers, animals, or, on occasion, even people.

To do scrollwork you need a jigsaw. Obviously a power jigsaw will speed your work considerably, although it's worth remembering that little of the scrollwork on nineteenth-century buildings was done with a power saw.

There are basically two kinds of scrollwork, or "gingerbread," as it's sometimes called: full-through and relief. The full-through scrolling involves just what the name implies—you cut all the way through the board you are going to use. For this, use lumber that is nominally an inch thick—1x4s, 1x6s, and so forth. Don't use plywood. Plywood is much harder to make clean holes in, and cut this way, even good exterior ply tends to come apart after a few seasons—or less.

For relief scrollwork, use strips of ¼-inch solid wood which will, after being cut out, be glued against a surface such as a wall or fasica board. Since the advent of plywood, it isn't easy to get ¼-inch solid wood, so don't be surprised if your lumberyard tells you there is no such thing. There is, but you may well have to get it through a cabinetmaker, who in turn can get it from one of the wood companies that supplies only the needs of cabinetmakers. If you find yourself utterly unable to come by ¼-inch wood, then use ½-inch tempered hardboard. It's not as durable in the weather as solid wood, but, if properly painted, it will last satisfactorily and it's easier to work with than plywood. It is obviously very easy to cut ¼-inch-thick wood or hardboard, so you have to work very carefully, particularly if you use a power saw. When you have your scrollwork all cut out, glue it to one-inch lumber, or to whatever surface it is you wish to decorate. Window and door cappings are usually of relief scrollwork. They pretty much have to be, if you aren't going to ventilate the wall!

Relief scrollwork, by the way, is particularly nice if you paint the background a different color from the scrolled piece. If you think that sounds interesting, paint the two pieces before you glue them together—you'll save yourself hours of the fiddliest painting imaginable, far worse than window frames.

Full-through scrollwork is most commonly used under eaves on the gable end of a building. To attach it, nail a strip of 1x2, wide side up, to the under edge of the eave, 2 to 3 inches in from the edge. Nail the scrollwork to the narrow side of the 1x2.

Decide on the type of scrollwork you want and then draw yourself a design on a piece of light cardboard, bought at an art store: good stuff and stiff, but not thick. For drawing utensils, you will want a compass, a ruler—preferably an architect's scale—and a french curve, all available at art stores.

Nr. 13.

This is a nice 19th century demonstration of lattice and scrollwork ornamentation. Taken as a whole it looks complicated and difficult, but the individual panels and decorative eave strips are repetitive and not difficult to do.

When you do your design, keep in mind the limitations of your scroll (or jig) saw. The distance between the bow (back) and the blade of a hand scroll saw is rarely more than 8 inches, which means that you can't conveniently handle a strip of board wider than that. Power jigsaws usually have a deeper throat, and can take a board from 12 to 18 inches.

If your design is to repeat—and most scrollwork is repeating—then you must obviously be sure that the two joining edges of the design (in other words, the beginning and the end) match up, line for line, hole for hole. One way to make sure of this is to decide the length of the design (say 18 inches) and then with a stencil draw in the first inch both at the beginning and *after* the 18th inch. Then fill in the rest of the design between.

Remember, too, to keep your angles in mind. Scroll designs usually have a true vertical and horizontal orientation, relative to the ground. this means that the angle at which they will be attached to the house must be taken into account when the scrollwork is done. Some designs, to be sure, look fine at any angle; others definitely do not. Pendant balls, for example, sticking out at a 45° angle to the ground look a little strange, while a flowing design of curves might look as good angled as horizontal or vertical.

When the drawing onto the cardboard is finished, cut it out with a model builder's knife (X-acto is one kind). You can now transfer the design to the wood you plan to use. The most obvious way to transfer the design is with a pencil or marker, using the cardboard as a stencil. That's fine, if you don't have very much stenciling to do. But if you have a lot of board to cover, you'll find the cardboard will begin to wear out, the holes getting larger, the angles less sharp. That means making a new stencil, and you may find it difficult to make a second one exactly like your first.

So I offer a solution: Do your stenciling with an air brush or, failing that, a can of spray paint. Air brushes are available from art stores, but they aren't cheap. On the other hand, they are better than cans of spray paint, because you can adjust the amount of paint down to the point where you can just make out the outlines you want, and you run minimum risk of soaking the cardboard. The can of spray paint is more generally available, of course, and lots

cheaper, but if you use one, be sure to follow the maker's directions as to spraying distance, or even increase it. You want if at all possible to avoid soaking the cardboard. If you have a very great deal of stenciling to do and you have to work with a spray can, melt some canning paraffin or some old candles and paint the stencil with the hot wax before you start to work. The wax will effectively protect the cardboard.

For the final touch in decoration, you can add a finial. Finials are the fancy solid ornamentations like the pawn in a standard chess set, that is, a sort of inverted funnel with ball at the top that frequently grace gable ends, roof peaks, or occasionally, turned-up rafter ends. They can also hang. Far more involved shapes may be used, though, and often have been. If you have a wood lathe big enough to turn a bedpost, then you can dream up all kinds of shapes for finials.

Fancy wrought-ironwork often took the place of both scrollwork and finials during the Victorian age, something you might like to look into. Check the Yellow Pages under Blacksmiths or Ornamental Ironwork. But beware—ironwork is expensive.

CHOOSING A GAZEBO DESIGN

A gazebo is a pleasant and rewarding thing to have in your garden, but if you build it yourself, the pleasure will, at the very least, be doubled. There are three reasons for this. First, building a garden house is a project in which you can give your imagination all the scope it wants. There is not, after all, a practical purpose that has to be served by the structure—it is purely a creature of fancy, built to serve fancy. Look at the pictures of gazebos in this chapter and consider the variety of styles people have come up with. You can make your structure open or closed, square or round, screened or glazed, modern or Victorian. And consider the choice of roof styles. The freedom is heady.

Don't feel constrained to build your gazebo in the same style as your house. A decade or two ago there was a widespread feeling that things adjacent to one another should all be of a single style; one did one's house in Danish modern or late colonial and wouldn't think of slipping in a piece of Victorian furniture, no matter how good it might be of its type. That has changed. And don't try to make your gazebo look functional; the modernist dictum that good function leads to good form is simply irrelevant to gazebos because they have no function in the practical sense—so let your fancy go.

One thing definitely not to be overlooked in building a gazebo is the possibility of using old materials. Most areas of the country with any population density have a used-lumber yard, often in conjunction with a house-wrecking company, where you may be able to find, in addition to the obvious used lumber, all kinds of decorative turned posts, wrought-iron work, finials, and scroll-work gingerbread—all potentially useful stuff for gazebo building. You might even be lucky enough to find the complete cupola of a demolished Victorian house. If you do, you've just about got your gazebo ready-built. Lay down a foundation and a floor, cut out a door, and you're in business. Quite a few gazebos have come about this way.

Although traditional gazebos are often round or polygonal in plan, rectangular garden houses and shelters have also been popular over the years. Unless you are a reasonably experienced builder, you will make fewer problems for yourself if you stick with the rectangular when you plan your gazebo. There is a lot of information about building small structures in the chapter about garden sheds, and that is where you should start when you plan the detailed construction. A purely utilitarian shed, however, with only one or two small windows, will not make a very satisfying gazebo, so you will almost certainly want to ring some fairly major changes on the basic themes. Windows, for example, can be made virtually as large as you like; be sure only that the headers and the sills and the studs on either side of the rough opening are all doubled. Studs that would normally be found where the window opening is should still be present, just interrupted by the opening. If you want a window to be wider than about 4 feet (interrupting two studs, if the studs are set on 16-inch centers), it's usually best to interrupt the window with a double stud, so giving you a double window.

If you mean the building to be only a *summer* house, you probably don't need glass windows; screens will be enough. Unless you can foresee a need to take the screens out at some time, it will be easiest if you simply stretch the screening over the window openings and fix it in place with screen molding held down with 1-inch finishing nails. Shutters, made like a board-and-batten door (see page 81) and hinged at the top, will protect the interior and the screening from the worst of winter, and when propped open in summer will keep most of the rain out. Once, when building shutters of that kind, I used fiberglass translucent awning panels instead of boards, in order to make the interior lighter with the shutters up or down. It was more expensive, but the desired end was achieved.

If you plan on a proper little house, with opening glass windows and the lot, you'll probably want to finish the interior walls. If it were a regular house you'd probably use plasterboard, but I'd advise against that in a gazebo, even a weathertight one. It is likely that some water *will* get in from time to time, and water is something plasterboard will not take. Instead, use one of the many plastic or plasticized panelings available, or ¼-inch hardboard, preferably tempered, and well enameled. If you want wallpaper, get some intended for use in a bathroom, and if the wallpaper is not self-adhesive, be sure your paste isn't water-soluble. If you like plain or varnished wood, remember that plain wood boards usually look a lot nicer than plywood.

If a round or polygonal gazebo is definitely what you want, it's worth noting that polygonal structures are easier to build than round ones, although both involve a number of peculiar construction problems not found in rectangular buildings. There are solutions to all these problems, but the solutions are not always easy to execute.

Round, then, or polygonal? And if polygonal, how many sides? One consideration that may influence your choice is the problem of windows. The openings present few difficulties so long as you don't want glass in them, and many delightful gazebos have windows that are simply screened and left unglazed. But if you want glass, you had best forget about a round structure; custom-rounded glass

This rustic gazebo is more interesting than it might have been because of the latticed windows and the diagonally-applied siding.

is prohibitively expensive. Settle instead on a polygonal gazebo with sides long enough to hold windows of the size you want.

Another consideration is ease of framing. The fewer the sides, the easier they will be to build. Round framing is the most difficult.

Below are instructions for the two more complicated types of gazebos—first for a polygon, then for a two-story round tower. Many of the construction features are common to both types of structure. The most important, the kingpost, deserves its own section.

Kingpost

Of central importance, both literally and figuratively, to any round or polygonal structure is the *kingpost*, the central "keystone" against which the upper ends of the rafters abut.

Basically, its function is to provide a surface to which the rafter ends can be nailed. As such, it does not need to be any taller than the rafters are deep. It may, however, be extended above the roof peak as an ornamental finial, if it is turned or otherwise decorated.

It may also extend down to the floor, or even through the floor to the ground or foundation. That way it helps bear the weight of the roof, and thereby does away with the need for stringers—the horizontal members that hold the lower ends of opposite rafters together and prevent the weight of the roof from bending the walls outward.

As a hexagonal structure has three pairs of facing walls, at least six stringers would be necessary—more if each wall segment were longer than four feet. The more sides to the structure, the more stringers necessary. So if uncluttered floor space is less important than an uncluttered-looking area under the roof, then using the kingpost as a central pillar in lieu of stringers could well be desirable.

The first rule about a kingpost is that it must present a flat side (although not necessarily at 90°) to each rafter end. This can be accomplished either by beveling the ends of the rafters to fit a square kingpost or by making a kingpost with as many sides as there are to be rafters, see Figs. 70 and 70a.

Here are two kingpost designs for you to consider. Fig. 71 on page 198 shows a kingpost made of three 2x6 boards glued and nailed together to form a square beam. (You may have to trim the width of the individual boards before gluing, to make the total width as small as the thickness; alternatively, you might add a 1x6 to the sandwich. It should be square in cross section.) It will support 12 rafters, radially as shown. Remove every second rafter, and you have the correct spacing for six rafters.

The difficulty with a square kingpost comes in cutting the ends of the rafters properly. They must be cut at a *compound angle*: you must mark the broad side of each rafter for the proper angle to get the slope of the roof (true for all rafters), and, for the rafters that don't meet the kingpost head-on, you must mark the edge with the appropriate angle (30° for six or twelve rafters). Once you've marked the angle, it's relatively easy to cut; if you're using any sort of a power saw, you can set the saw for the 30° bevel and saw across the face marked for the other angle.

The Fig. 72 on page 199 shows a different type of kingpost made of three 2x6 boards glued together, but shaped to support eight or sixteen rafters. To make this kind of kingpost, first trim the boards so they'll form a square beam, glue them together, clamp them well, and let the glue set thoroughly. The width for each of the eight sides will be about 2 inches. Find the middle 2 inches of each full side (one inch either side of the center point) and mark them. Diagonals drawn between marks on

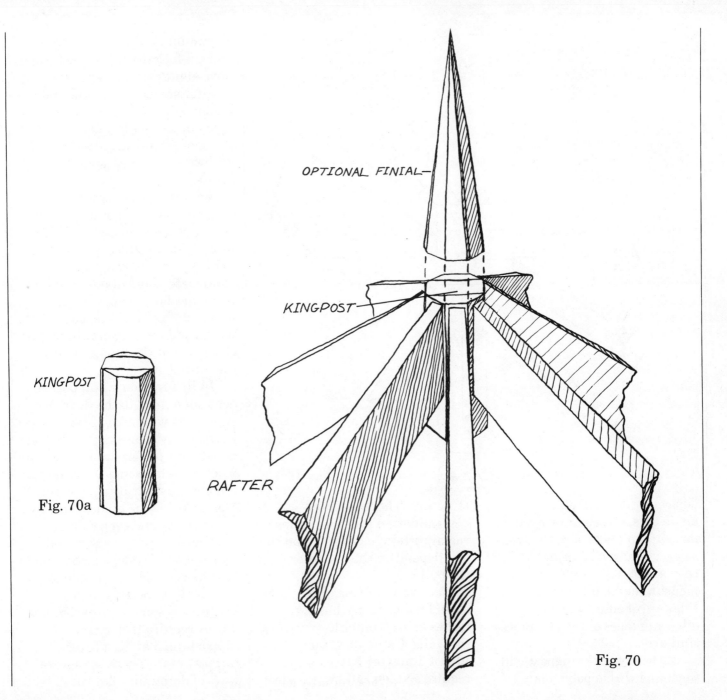

OPTIONAL FINIAL—

KINGPOST

KINGPOST

Fig. 70a

RAFTER

Fig. 70

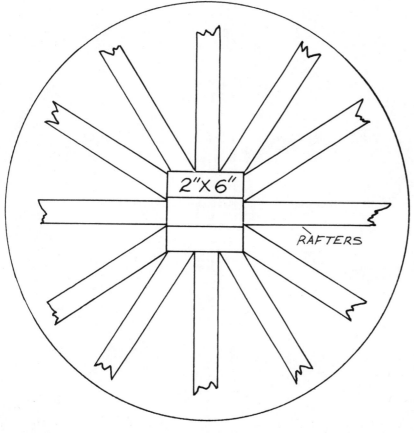

2"X6"

RAFTERS

SQUARE KINGPOST   Fig. 71

one would need twelve rafters, and a 6-foot shelter would have to have eight rafters. (The two smaller structures would find the rafters slightly more than 2 feet apart at the walls, but because of the short spans involved, that is still well within acceptable limits.)

For other sizes, whether the structure is round or polygonal, multiply the diameter *in inches* by 3.14 (pi) for the circumference (which you'll need for figuring siding, etc.), and divide that by 24 inches for the appropriate number of rafters. It is possible to use an odd number of rafters, but easier to cut the kingpost with an even number.

In its function a kingpost that does not extend to the floor as a support is very much like a keystone in a masonry arch, and getting it up there in the first place presents some of the same problems as setting a keystone. Mortarless stone arches built by the Romans and even earlier civilizations were constructed over wooden forms, which could be removed once the keystone was in place. Setting up a kingpost is somewhat easier, but it does require temporary support until at least three rafters are in. There are several ways of doing this, but the

adjacent sides will place the angle cuts, but before you cut, measure to make sure the angled sides will also be 2 inches wide. If they aren't, adjust all your measurements until the eight sides are of equal width. For other numbers of rafters, proceed similarly.

Either type of kingpost will work fine with a polygonal structure. The important determination is of the number of radial rafters to be used, and that will depend on the diameter of the building.

As a general rule, the rafters should not be more than 24 inches apart, center to center, at the walls. Using this rule, a 10-foot diameter gazebo would require sixteen rafters, an 8-foot method I've used was to toenail a 2x4 to the bottom of the kingpost, long enough to reach the floor when the kingpost is at the right altitude. Before sticking the kingpost up in the air, though, I nailed one rafter to its side, giving me a wooden composition that looked like an anchor missing one arm. To keep it all from getting out of shape, and nails from pulling out, I needed help in setting it up, and arranging the rafter so it rested properly on a wall plate. Once that was done, though, I was home free. The next rafter was the one diametrically opposite the one already in place. With two rafters firmly nailed both to the plates and the kingpost, the rest presented no problems. When four rafters set 90° apart, were in, I removed the temporary prop from the kingpost.

If the kingpost doesn't reach the floor, you will need stringers to keep the weight of the roof from forcing the walls apart. In a small round or octagonal building, 12-feet in diameter or less, four stringers are enough, arranged in a Greek-cross pattern, like the framework of a tic-tac-toe game. In an octagonal building, one stringer end is nailed at each corner; in other

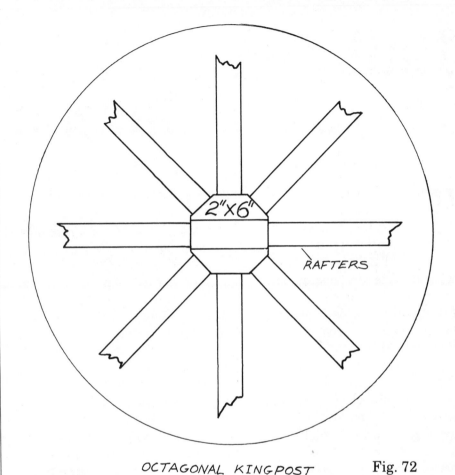

OCTAGONAL KINGPOST   Fig. 72

polygonal or round structures nail the stringer ends at points equidistant around the circumference. Although it is usual to stand stringers on their narrow sides and nail them to the sides of rafters, where both lie on the plate, it is perfectly acceptable to lay the stringer on its wide side, and nail it to the wall plate. Mind you, that must

be done before any roofing is laid on, or you'll have no room to swing the hammer! Where the stringers cross, two will have to bend up a little and two down, but that's okay. If you want to be fancy, you can arrange them in a woven pattern, each one first crossing over, then crossing under a 90° mate.

Building a Polygonal Gazebo

T here are some general guidelines for building a polygonal gazebo.

First of all you must decide on its general form and size; whether it is to be hexagonal, octagonal, or whatever, have fully open or partially open sides, be screened or glazed, and what its diameter is to be.

The best foundation will depend on the size of your structure, and the terrain on which it is to be built. A gazebo of average size (8 to 12 feet in diameter) on firm, level ground needs no more than a concrete block at each corner, and one in the center, if you use radial joists (see below for joist options). Be sure the blocks are level with one another by laying a straight board from block to block and checking it with a level.

Softer or more uneven ground will require a more elaborate foundation. Unless your ground is unusually soft, it will normally be sufficient to pour a concrete footing for the concrete block. See page 90 for help in that. Very uneven ground may require one or more posts. For help in setting a post, see page 83.

For any polygonal gazebo up to 12 feet in diameter, 2x8 is more than sufficiently sturdy for the perimeter boards.

Here are two ways to arrange the floor joists. If you do not plan on a full-length, bearing kingpost, easiest and best is to treat the polygonal floor as though it were rectangular. Make all the joists parallel to one side (or two, if the building has an even number of sides), and saw them as needed to fit perimeter boards they meet at an angle. Use 2x6 joists and 2x2 joist carriers.

If you want to avoid stringers, and so use a kingpost bearing on the floor, best to use radial joists, supported at the center by a footing just like the ones you made at the perimeter. Because the unsupported spans are thus halved, you can safely use 2x4 joists, if you wish. Run one joist all the way across the building (a *full joist*) and then fill in additional joists half as long (*cripple joists*) nailed at the center to the full joist. Start with the cripple joists at right angles to the full joist, then progressively bisect the angles, until the ends of the joists are no more than 2 feet apart at the perimeter. Toenail the inner ends of the cripple joists, and be sure they are all resting solidly on the central footing. The central footing will also, of course, support the kingpost.

FRAMING

Framing the walls and the opening for doors and windows on a polygonal gazebo is no trick. Use standard house-frame construction methods (see the chapter on garden sheds), with one notable difference: at each corner use two 2x4 studs. In cross section they should form a V, with the inner edges touching but separated at the outside edges so that each 2x4 presents a flat edge to its side of the structure, Fig. 73.

You may, in fact, want no walls at all, just pillars supporting the roof. For this

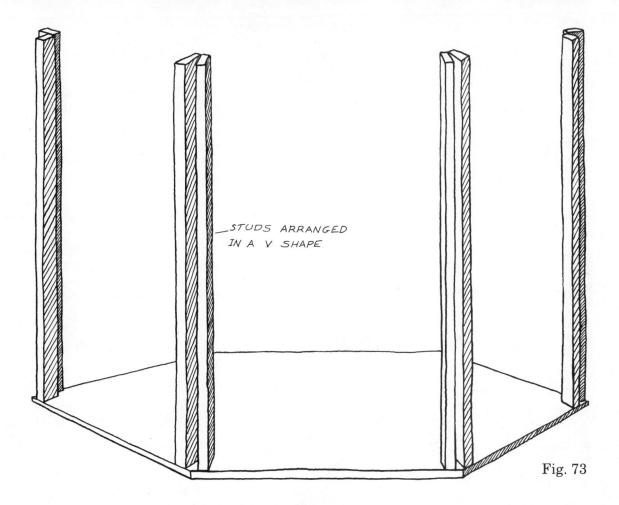

STUDS ARRANGED
IN A V SHAPE

Fig. 73

purpose, two 2x4s nailed flat together make a tidy post, or you might like to use ready-built pillars, which are available in varying diameters from your lumberyard. If you want open sides with pillars, you should do away with shoe boards, as people stepping into and out of the gazebo will regularly be tripped up by them. Simply toenail the posts or pillars to the floor.

RAILING

You don't *need* a railing if you go for open sides, but it can look nice, and it is pleasant to rest your feet on when you're sitting inside. A good basic railing is easy to make out of good-quality, kiln-dried 2x4s. Don't use cheap, undried or unseasoned wood for a railing, as it will warp. Simply cut lengths of 2x4 to fit between the pillars and toenail them in place at waist height.

Nicer, but a little more work, is to support the ends neatly. Cut an 8-inch-long piece of 2x4, then

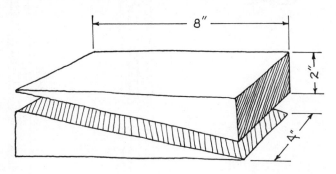

Fig. 74

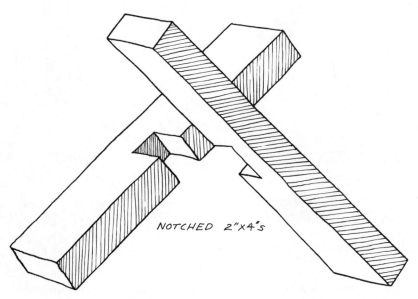

NOTCHED 2"x4's

Fig. 75

saw it diagonally, as shown in Fig. 74. Nail the 8-inch side to the post with the thick end of the wedge supporting the railing. This works only with square posts, of course.

If the unsupported run of a section of railing is more than about three feet, you may want to support it with a post. In that case, use a shoe board. You can either prefabricate the "wall" unit, consisting of shoe, pillars, railing and post, and plate, or you can toenail your posts or pillars to the floor, and then cut shoes to run between them. Nail the post(s) onto the shoe, nailing up through the underside of the shoe, then nail the shoe in place on the floor. Then put the wedge-shaped railing supports on the pillars, and nail on the railing.

Once the basic railing is in place, there are a number of things you can do to dress it up. A simple X-frame of 2x4s looks nice in each opening. Notch the two boards so they fit together where they cross, as in Fig. 75.

You can also fill in the openings with plywood, either inside the X-frame or outside it. Nail strips of 1x2 to the underside of the railing, the sides of the pillars, and the floor. The inside edge of the 1x2 should be flush with the inside edge of the railing. Cut the plywood to fit inside the railing opening, against the outside edge of the 1x2, to which the plywood can then be nailed.

Lattice also looks nice in the railing openings and is simple to make, although fairly expensive. Lattice strips come in a variety of widths, from one inch to about three inches. A width of 1½ inch looks nice, but if you like wider lattice, you're lucky, as the job will cost less.

To figure out how much lattice to buy, calculate the area of the spaces to be filled. If your railing is 3 feet high and the pillars are 3 feet apart, the area to be filled is 9 square feet. You will need lattice of exactly the same area, assuming you'll be using the standard crossed-lattice formation. So figure the area of the lattice you'll use; if you choose 2-inch-wide lattice, a strip 6 feet long will have an area of one square foot (or 144 square inches: 6 feet x 12 inches per foot x 2 inches wide = 144 square inches). So for your 3x3-foot opening, you need nine times 6 feet, or 54 running feet of 2-inch lattice. Add another 6 feet for waste. Get strips as long as you can manage

easily—say 10 or 12 feet—to reduce the number of unusable short ends. Some of the short ends can be used in the corners.

Lattice can't be toenailed—it's too thin—so you must arrange something to nail it to. Strips of 1x2 nailed wide-side-down all around the inner edge of the opening will do nicely as nailers. You can either space the 1x2 so that the lath is just flush with the outer surface of the building, or so that it's recessed. Use ¾-inch brads to nail the lattice to the nailers, or a staple gun with ½-inch staples—longer, if your staple gun will accommodate them.

Two things are critical for doing a lattice job right. The ends of the strips must be cut at an even angle (45° is usual), and the first strip must be nailed on with great care, so its angle to the railing or pillar is right (again, 45° is usual). Use a miter box for cutting the end angles, and when laying on the first strip, measure at least twice to be sure the vertical and horizontal distances from the corner to the ends of the strip are the same. Don't start right up in the corner. Start with a strip that extends from an upper corner to the bottom of the area to be filled.

Next, lay on all the strips that

run in the same direction as the first strip. Each strip should lie exactly the width of a lattice strip from its neighbors—that is, if you use 2-inch lattice, leave 2-inch spaces between each pair of strips. Do that top and bottom before nailing, and you can't go wrong. To complete the lattice, you simply repeat the process with strips running at 90° to the first course. Set the second course so that their ends lie exactly on the ends of the first course, or your edges will look a little ragged.

You will probably find, particularly after a few weeks have passed, that some of the strips have bowed in or out a little. If this bothers you, use a staple gun filled with ceiling-tile staples to fasten crossing lattice strips together. Bend the staples over on the inside.

ROOF

After the studs and plates of the framing are in place, the next step is to put up the rafters, kingpost, and stringers, as described in the section on kingposts. Once you have the rafters and stringers in place, you have to put up the underroof. Basically, a faceted roof is easy: simply cut a separate triangle of ⅝-inch plywood to cover the space between each pair of rafters. Measure carefully, from center to center of each rafter; slight differences are likely to have crept in. If you can manage it, bevel the cuts so the two adjoining wedges of plywood will meet neatly. More details are given on page 218.

Even if you're planning on an open-sided gazebo, it's a good idea to have a tight roof, both because moisture trapped between the roof sheathing and the rafters encourages mildew and eventually rot, and because it's irritating to have a drip coming from the ceiling for fifteen minutes after the rain has stopped. Normally shingles over an underroof are quite enough, but because of the angular breaks in the roof of a polygonal gazebo, it's a little harder to be sure the shingles are sealing tightly. Although it won't keep water from between the shingles and the underroof, you can at least keep it from coming all the way through by laying a bead of caulking compound along each rafter edge before you put down the pieces of underroof. If you think water is—or might be—getting through the shingles, you can nail a strip of aluminum or galvanized flashing, about 4 inches wide, down each seam, on top of the shingles.

The easiest finish roofing to apply is composition shingles. You'll want to overlap the edges of the shingles at the seams between the roof planes, and run a row of single shingles up the seam. Good directions for applying shingles are found on the wrapper of each bundle you buy.

EMBELLISHMENTS

Scrollwork is a traditional embellishment of gazebos and would look nice under the railings, along the posts, and under the eaves. For directions, see the information on scrollwork on page 191.

At the peak of the roof the nicest touch is a finial capping the kingpost, or perhaps a decorative weathervane. If you go this way, you must apply and caulk a strip of metal flashing (either aluminum or galvanized steel) around the finial or the weathervane, and overlapping by at least 4 inches the top course of shingles.

An easy if not quite so decorative alternative is a simple metal cap. Cut a 2-foot circle out of sheet aluminum, of the kind hardware stores sell for do-it-yourself projects. Make a slit in the circle from the center to the edge. Place the circle on

the peak of the roof and, sliding one edge of the slit over the other, make a shallow cone of the metal that just fits the curve and pitch of the roof. Caulk the seam and nail it in place.

When you are finished with whatever ornamentation you decide on, consider painting it either a color contrasting to that of the rest of the gazebo or a slightly different—best, a lighter—shade of the same color. Such treatment will emphasize the trim and bring out the shapes and forms of the structure.

Building a Two-Story Round Gazebo

R ecalling the putative pseudo-Latin etymology of *gazebo*—"I shall gaze"—the assumption, of course, is that a gazebo is properly something from which you gaze out over a fine vista. With that bit of folk etymology in mind, here is a terrific plan for a gazebo at the top of a small tower. The bottom part can be anything you want it to be—garden shed, playhouse, poolside cabana, outdoor kitchen, anything. The top part is a screened-in (or glazed, if you like) gazebo.

This tower, Fig. 76, is 8 feet in diameter and about 20 feet 5 inches high, not including a finial on top of the roof. The headroom downstairs measure 7½ feet; upstairs it measures 7 feet even.

FOUNDATION

The tower needs a proper poured-concrete perimeter foundation, 8 inches thick and 24 inches deep—16 inches underground

Materials Needed

about 40 stakes, made of 1x2 furring strips, or similar scrap. See text, step 2 for how to figure the length.

one 4x8 foot sheet of ¼-inch plywood, cheapest grade, for concrete forms

1-inch box nails for concrete forms *or* ½-inch staples in a stapling gun

eighteen ⅜-inch carriage bolts, 10 inches long, with washers and nuts, to secure floor to foundation

about 1¼ cubic yards of concrete

four foundation vents *or* ordinary concrete blocks (with holes)

Construction Steps

1. Drive a short stake into the ground in the center of the circle your building is eventually to occupy. Drive a nail into the top—big enough to tie some stout string to or twist a wire around, but not so fat as to split the stake. Use the wire or string to locate two concentric circles of stakes marking the inner and outer edges of the foundation, see Fig. 77, page 207. The inner circle should have a radius (distance from the center post) of 3 feet 4 inches; the outer circle needs a radius of 4 feet. If you locate the stakes about 14 inches apart on the inner circle and 16 inches apart on the outer one, they will be about evenly spaced. You don't need to drive the stakes in very deeply; you only need mark the inner and outer edges of the foundation trench.

2. Dig the trench 16 inches deep. How you proceed with the next step will depend to some degree on the sort of soil you're building on. The trench will contain the bottom two-thirds of the foundation, but the upper third extends 8 inches above the ground. You must therefore extend the trench upwards, and that is done by building temporary walls of thin plywood. The plywood will not stay in place by itself, so it will have to be held with stakes, from behind. As ideally the inner sides of the

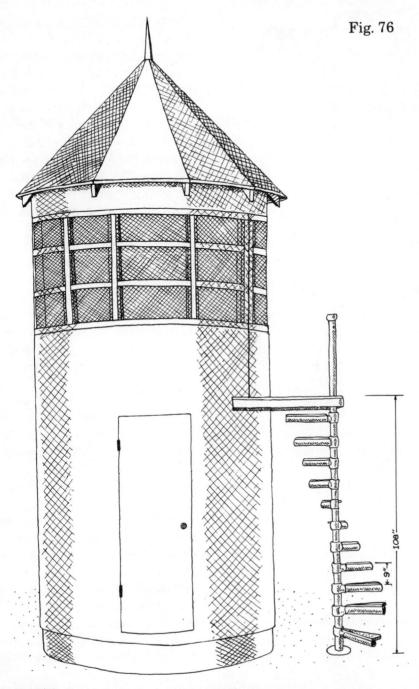

Fig. 76

plywood form will be even with the edges of the trench, the stakes will need to be driven into the ground just outside the edges of the trench. If your soil is firm but not crumbly, you may be able to back the plywood adequately with stakes driven 8 or 10 inches into the ground, just beyond the edges of the trench, without disturbing the trench walls. If so, you need stakes only about 18 inches long. More likely, you will have to recess the stakes in the walls of the trench, ideally set back from flush with the trench walls by ¼ inch—the thickness of the plywood. In that case, drive the stakes into the bottom of the trench to a sufficient depth so that the stake firmly resists being pressed sideways. You may have to experiment a little before you can determine the length stakes you need. Remember, 8 inches must protrude above ground to back the plywood.

If rocks or excessively soft soil make the foregoing impossible, you may have to widen the upper third of the foundation, so you can set the stakes four or six inches from the trench edges, giving you a foundation with a "T" cross-section. If your soil is so friable or loose as to require this course, the extra foundation width will probably be a good

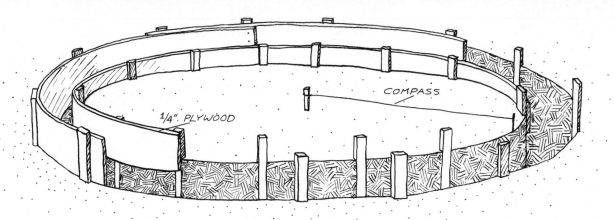

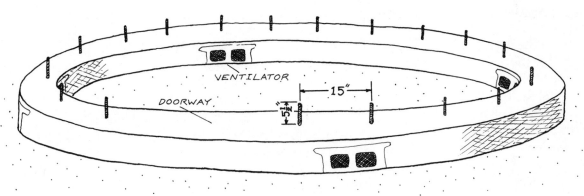

Fig. 77

Fig. 78

thing in any case.

3. Cut a 4x8-foot sheet of cheapest grade ¼-inch plywood into six strips 8 inches wide and 8 feet long. Using a stapling gun and ½-inch staples or a hammer and 1-inch box nails, fix the strips to the outer faces of the inner ring of stakes, and to the inner faces of the outer ring of stakes. Take great care to make the upper edges of the form precisely level, as it will determine the top surface of the concrete, on which the building will be built. Check the form by laying a straight 2x4 all the way across it, and placing a level on it. Rotate the 2x4, checking every few feet all the way around. Remember: If the form isn't level all the way around, the top of the foundation won't be level—and you will have something more like the leaning tower of Pisa than you probably want.

4. Fill the trench up to ground level with concrete—*not* to the top of the plywood. Screed it well with a hoe handle or the like, to get all the air bubbles out. At 90° intervals—at 3, 6, 9, and 12 o'clock—place four foundation vents into the form. These vents may be bought from a builder's supply company, or you may use 8-inch-thick concrete blocks, set so the holes are horizontal. Don't worry if they press the form out of shape a little.

5. When the vents are in place, fill the rest of the way to the top of the plywood form with concrete, see Fig. 78, page 207. You will need about 1¼ cubic yards of concrete for the whole foundation. As most ready-mixed concrete companies will deliver any quantity of a cubic yard or more, I suggest you get it that way, if you can. It's much the easiest, and not that much more expensive than mixing it yourself.

6. Now place the bolts, head down, into the concrete. Space them evenly every 15 inches around the foundation, centered between the edges, leave a gap where your doorway is to be. Each bolt should protrude 5½ inches from the concrete, and

must be vertical. Check them with a square every half hour or so until the concrete has set to the point where the bolts can no longer move.

7. When the concrete has set, remove the form.

FLOORS AND WALLS

The most unusual parts of this round construction are three heavy, stiff disks, each resembling a sandwich, as shown in Figs. 79 and 80, on page 209. The framing is similar to normal house framing, with some differences due to the found form. The siding is a double layer of thin plywood, flexible enough to curve to the needed shape. Since the skin (both of the floors and walls) is important structurally, glue must be used, and everything is built in a single process, with part of the siding going on before the second floor is completed.

Materials Needed

Lumber for the Disks:
six 4x8-foot sheets of ¾-inch plywood, exterior grade, one side good
six 4x8-foot sheets of ½-inch plywood, exterior grade, one side good
sixteen 8-foot 2x4s for disk joists

three 12-foot 2x4 for second-story joist and landing support
twenty 8-foot 2x4s (or the equivalent in scrap) for 4-inch spacer blocks

Lumber for the Walls and Doors:
forty-one 7½-foot 2x4 studs
four 7½-foot 2x4 verticals for doors
three 7½-foot 2x4s for first-story horizontal braces.
three 7-foot 2x6s for second-story horizontal braces
one 7-foot and one 10-foot 2x6 for horizontals for doors
twenty-six 4x8-foot sheets of ¼-inch plywood, exterior grade, one side good, for siding (including doors)

Other Materials:
4 pounds 2-inch serrated siding nails
12 pounds 3½-inch common nails for framing
5 pounds 4½-inch nails for framing (as called for)
1200 2-inch #6 flathead screws for siding
2 gallons carpenter's glue plus brush

Construction Steps
The heart of this round construction is found in three disks, each of which is like a sandwich, with a top layer of

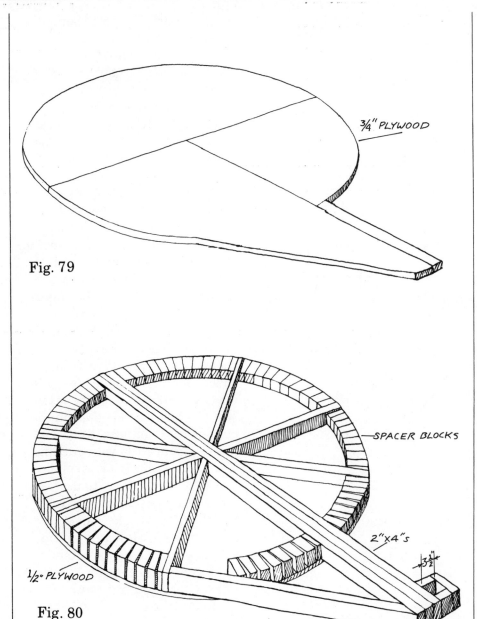

¾″ PLYWOOD

Fig. 79

SPACER BLOCKS

2″X4″S

3½″

½″ PLYWOOD

Fig. 80

¾-inch plywood and a bottom layer of ½-inch plywood. In between are the radial joists and a lot of 4-inch 2x4 spacer blocks. See Fig. 80. The spacer blocks around the edge serve two purposes—they contribute to rigidity and they provide substance into which the siding and the studs can be nailed.

Fig. 79 shows the second-story disk, in which the full-length, tripled joist is additionally lengthened to help support the landing for the stairs. The first-floor disk and the disk under the roof are simple circles. Each circle of plywood must be cut from two 4x8-foot sheets of plywood, so there will be a seam in the middle. Note that these seams are at 45° angles to the full-diameter joist. The one exception is the upper surface of the second-floor disk. For it you must cut one semi-circle, one quarter-circle, and another quarter with an added triangle, as shown in Fig. 79. Those parts go together as shown in the main disk illustration.

1. While waiting for the foundation to set, you can start on the disks. Lay two sheets of plywood of the same thickness next to one another, to form an

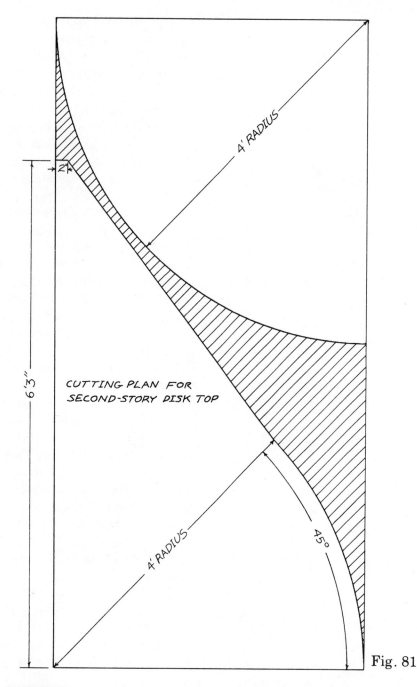

CUTTING PLAN FOR
SECOND-STORY DISK TOP

4' RADIUS

4' RADIUS

45°

6'3"

2"

Fig. 81

8-foot square. Support them with three 8-foot (or longer) 2x4s, running at right angles to the seam between the two sheets. Exactly in the middle of the square, drive a thin (box or finishing) nail through the crack between the sheets into the 2x4 below. Leave enough of the nail standing proud to wrap one end of about 4½ feet of thin wire around. Wrap the other end of the wire around a pencil, just above the point, so that the point is exactly 4 feet from the nail, that is—just able to reach the center of each edge of the square. Using the pencil and wire as a compass, draw a full circle on the plywood, thus marking two sheets for cutting. Five ¾-inch sheets should be so marked, and six ½-inch sheets. Cut the sixth ¾-inch sheet as indicated on the cutting diagram, Fig. 81. Use a power jigsaw to do your cutting, if you can. If electricity isn't available, you'll have to use a hand keyhole or saber saw, which will take a lot longer. It may be worthwhile taking the sheets to the nearest electric outlet and, if necessary, buying or renting a power jigsaw.

2. Put the first-story disk together. All the joist cuts are 90° except the inner ends of the four radial half-joists, which have 45°

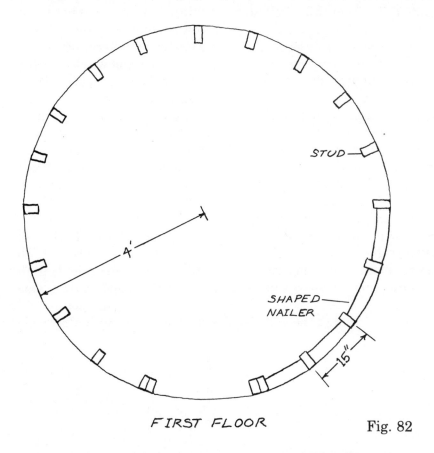

STUD

SHAPED
NAILER

4'

15"

FIRST FLOOR   Fig. 82

cuts. Glue the triple joist together, then nail it together with 4½-inch nails. Space the nails about every 8 inches, and you won't need to clamp. Then lay the joists out on a flat area, toenail them together at the center, and lay a double bead of glue along the upper surface of each joist. Lay a plywood semi-circle in place, with its flat edge at a 45°angle to the main

joist. Tack the edge of the plywood to the ends of the joists after measuring their spacing, starting with the main joist and going both ways. Center to center, the joist ends should be 37¾ inches apart, *measuring around the curve of the plywood*.

Before the glue dries, lay on the other semi-circle, and tack it to its joists, after measuring. Check all the spacings again, and

then nail it all firmly in place with 3-inch finishing nails set about 8 inches apart along each joist.

Turn the disk over and apply the other two plywood semi-circles with glue and nails. Be sure you have ¾-inch ply on one side and ½-inch ply on the other. The seam on the second side should be at right angles to the seam on the first, and 45° from the main joist.

3. Cut the 4-inch long 2x4 spacer blocks. You'll need about 150 of them for each disk; a few less for the middle disk. Cut only what you need for the first disk, however, as you will be getting some scrap ends of 2x4 as you build. Lay a double bead of glue on each block, top and bottom, and slide it into place. The inner ends of the blocks should touch. Nail each block in place with three 2½-inch nails: two through the ¾-inch ply into the block near its ends, and one through the ½-inch ply into middle of the block. Don't use larger nails, lest you split the blocks. Be sure that the outer end of each block is just flush with the edges of the plywood disks.

4. When the foundation has set solidly (give it at least 24 hours of dry weather), place the disk onto the upturned bolts of the foundation, with the ½-inch plywood down. Center it accurately. Then, just barely lifting the disk, place a little piece of ordinary carbon paper, carbon side upwards, between each bolt end and the disk. Make sure the disk doesn't shift, as you strike it over each bolt end once with a hammer, hard enough so the bolt end and carbon paper leave a mark on the underside of the disk. Remove the disk and drill ½-inch holes straight through at each point marked by a bolt end. Set the disk in place and draw it down with nuts. Use a flat washer under each nut.

5. The wall framing of the first story is done much as in any frame structure (see the chapter on garden sheds for basic directions), but there are obvious adaptations. The shoes and plates, for example, are replaced by the disks, and all other horizontal members are also curved.

Cut the studs 7½ feet long. If you are able to buy cheap 7½-foot studs from a discount lumber-yard, you may find some of them are a little short. If so, it won't hurt a thing if you trim *all* the studs to 7 feet 5 inches or even 7 feet 4 inches, as long as all the studs are identical in length. If you plan no windows on the

ground floor, you will need 21 studs, to be set on 15-inch centers, with the studs on either side of the 30-inch doorway (rough opening size) doubled, Fig. 82, page 211. If you plan windows, adjust the number of studs accordingly. The circumference of the disk is just about 300 inches, which is why 15-inch stud placement and a 30-inch door were decided upon—they divide evenly into 300. (Keep that in mind if you must resite studs for windows. Remember that the horizontal window-frame members must be curved to conform to the building, although the window

units themselves, if not more than 30 inches wide, can be flat. The upper surfaces of the rough-opening window sills must be exactly 35¼ inches above the floor).

Toenail each stud to the disk, using one 3 ½-inch nail on each side of each stud. You'd better not do this on a windy day, or you will have difficulty keeping the studs upright.

6. Start on the second disk, fastening the joists to the ½-inch plywood semi-circles. Note that the main, tripled joist is 10 feet 3 inches long, and the tangential joist is about four feet long; measure and cut it after the

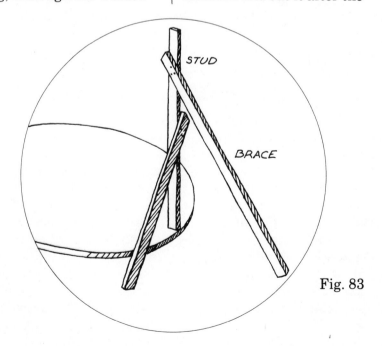

STUD

BRACE

Fig. 83

other joists are in place. Leave off the top ¾-inch plywood disk and the spacer blocks, but finish-nail the ½-inch ply and the joists together firmly. Don't forget the glue!

7. The studs, secured only with a couple of toenails to the floor, will need bracing before you lay the second disk on top of them—which is the next step. No need to brace all of them, though. Four evenly spaced around the circle are enough. Each of those studs should have one brace extending to the ground directly out and away from the tower, on what would be an extension of a radius of the circle, and another brace extending to the ground at 90° to the first brace, in line with what would be a tangent to the circle see Fig. 83. Use 2x4s for the braces, long enough to meet the ground at least 4 feet horizontally from the base of the stud being braced. Before staking the braces to the ground, check with a level to make sure the braced studs are perfectly vertical.

8. Lay the partly completed disk, plywood side down, on top of the studs. You will probably need some help with this step, for you must be careful not to push any of the studs to the side as the disk is laid in place. Also, to avoid

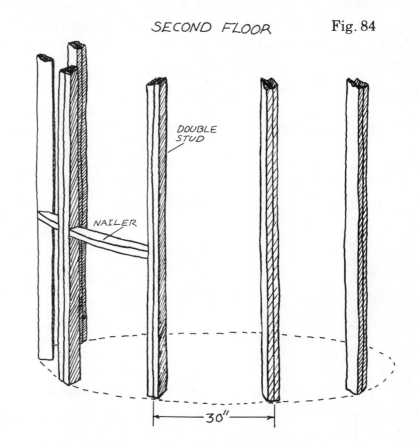

SECOND FLOOR   Fig. 84

DOUBLE STUD

NAILER

|←—— 30″ ——→|

twisting the studs away from vertical, stand on a stepladder—do not kneel on the disk itself—when nailing it in place to the tops of the studs. Use 2½-inch nails, two for each stud. Be sure each stud is vertical and just where it should be before you nail the disk to it.

9. Next it's time to cut, shape, and put in place the horizontal nailers on the first floor level. Their centers (not their upper

surfaces, as with the window sills, if any) must be exactly 36¾ inches above the floor. Make the nailers from 2x4s. If you have placed your studs 15 inches apart, each nailer will be about 13½ inches long, but be sure to measure each space before cutting the nailer. They must be shaped as shown in Fig. 84. The arc of the outer edge has, of course, a 4-foot radius. Use a piece of the ½-inch plywood scrap from the semi-circles to mark the curve onto the nailers. The end cuts will not be at 90°, but at about 82°. Your cutting will most easily be done if you have access to a bandsaw. Failing that, you can clamp the nailers to the top of a sawhorse and cut with a power jigsaw. If you must use hand tools, don't use a saw at all; use either a plane or a Surform, with the wood held firmly in a vise. There is, at least, great scope here for satisfaction if you like to do small jobs well, or admire and wish to emulate the craftsmanship of pre-power-tool woodworkers.

You can speed up the accurate installation of these nailers by cutting a pair of boards or timbers (whatever's handy) exactly 36 inches long. Place these boards against two adjacent studs, and lay the nailer on top of them, and nail it into place. Do that for each nailer, and you may be sure they will all be at the right height.

10. When the nailers are all in place, complete the second-story disk by gluing and nailing in place the spacer blocks and the top segments of ¾-inch plywood. You will notice the outer timber of the triple main joist and the box frame at the end of the landing are not covered by the plywood. To improve appearance and to provide better footing, cut from some of your scrap a strip of plywood that will cover the joist, but don't cover the box frame.

11. For the sake of structural rigidity, it is a good idea now to lay on the first course of siding. The best siding for this circular tower is ¼-inch exterior-grade plywood, good one side (grade A-C). The siding will be put on in two layers. The first layer will have the good side facing inward; the second layer, the good side facing outward. The first course of siding will consist of three plywood sheets standing on their sides, each stretching about a third of the way around the tower. Each will be trimmed to be 41½ inches high and about 90 inches long. The 41½ inches must be exact; the 90 inches adjusted so the ends of the sheets all fall right in the middle of a stud—except at the door, where the plywood should cover the whole width of the studs framing the door rough opening.

Start at one side of the door opening. The top of the plywood must be at the centerline of the nailers, and flush with the tops of any window sills. The bottom of the plywood should just cover the bottom edge of the lower disk. If it overlaps a little, don't worry. Paint glue on the disk, stud, and nailer surfaces for about a foot from the edge of the sheet at the door opening. Then tack the end

FIRST-STORY OPTIONS

The lower section of the tower could make an excellent dovecote. To make it that, you have only to cut 2½- to 2-inch holes in the upper wall of the downstairs section, about 12 inches apart, preferably staggered in two horizontal rows about 9 inches apart, with about 20 inches between holes on each row. Then, on the inside, affix foot-square platforms to the wall, about 4 inches below each hole. It will help birds if you then put a little edge on each platform made out of lath or lattice strips.

The lower floor could also be used as a tool shed, or as a poolside changing room. With the addition of a couple of windows, it would make a good children's playroom, or possibly a private place for someone in the family to work.

of the sheet in place, using three or four nails pounded only partway in. Bend the sheet around, then, to make sure it's horizontal. The upper edge should meet the centerline of the nailers. When you're satisfied with the position, drill and set the top and bottom screws. Use 2-inch #6 flathead screws, set 4 inches apart. Remove the temporary nails, and when you have set in all eleven screws, bend the sheet around the tower until you can nail it to the next stud, using 2-inch serrated or threaded siding nails, set every 6 inches. Paint glue on the disk, stud, and nailer surfaces to be covered by the sheet as you wrap it around. When half the sheet is firmly nailed, bend the rest of it around to mark for its cut. You'll want to end right on the centerline of the seventh stud. Be very careful to make the cut precisely vertical, as the next sheet will have to butt up against it. Nail and glue the rest of the sheet in place, then, until you reach the edge, which you must screw to the stud as you did the first edge. Apply the other two sheets in the same way.

12. Now it's time to frame the second story. Here, instead of individual studs set 15 inches apart, use doubled studs set on 30-inch centers. This will permit larger window openings. Toenail the studs in place directly over alternate first-story studs. Place the first double stud directly over the main triple joist, where it extends out over the landing, and proceed around the disk from there. If your weather is at all windy, by the way, you might have better luck keeping your studs in position if you set in the shaped nailers at the same time (see step 13, below). Do not place a nailer opposite the landing extension, as that is where the second-story door will be.

Use 7-foot studs on the second floor. If you saw them from 7½-foot precut studs, you'll have some of the spacer blocks you'll need for the third disk.

13. The shaped nailers on the second story must be set so that their upper surfaces (not centerlines) are exactly 38 inches from the floor. The nailers are, of course, cut essentially like the first-story nailers (step 9), but because they are longer, the curve is deeper, so you'll have to use 2x6s instead of 2x4s. The end cuts will be at about 74°. Be very careful that the two adjacent studs are vertical before you measure between them for nailers. You will minimize risk of error if you measure at the bottoms, where they are nailed to the disk. This, of course, assuming you have straight studs! Remember not to set in a nailer where the door is to be, at the landing.

14. When you have the second-story studs and nailers in place, it's time to complete the siding. For the next course, use 8-foot sheets standing on end, cut about 30 inches wide, so that each sheet exactly spans the spaces between three studs downstairs, and two studs upstairs. As before, start at one side of the downstairs doorway. Attach the sheets just as you did the first course; first tack-nail an edge in place, check for positioning, then paint the first surfaces to be wrapped with glue, and screw the edge of the siding sheet into place. Use the same screws, set 4 inches apart. Trim the second edge as you wrap. The bottom of this course of siding should rest on the top of the first course, and extend exactly to the top of the second-story nailers.

The one tricky part of the second course of siding is trimming one or two (depending on whether you started to the left or the right of the downstairs doorway) of the sheets to fit around the two protruding joists for the landing. You want the siding to cover completely the doubled stud at the right side of the upstairs door opening (seen from outside), then cut straight down to the bottom of the main joist, cut horizontally under the main joist, then cut up to the upper surface of the disk, then horizontally to the left as far as the next stud, and up it to the upper surface of the nailer. A rectangular cut-out for the tangential joist will have to be taken out of the side of that sheet of the siding. The trick to doing these cuts is not in the cutting—but in the measuring. Measure at least twice for each cut, and don't cut anything until you have drawn the whole pattern out onto the sheet of siding. That will give you an idea if you have things right. Double check everything. If you don't, you're all too likely to have to throw away a piece of plywood and start in again.

15. The second layer of siding should be laid on in a pattern upside down and staggered from that of the first layer, with the vertical sheets at the bottom and the horizontal ones at the top. The tops of the 8-foot-high, approximately 31-inch-wide lower course sheets should reach exactly 1¼ inch past the bottom edge of the second-story disk;

their lower edges should just meet or slightly overlap the bottom edge of the bottom disk. The second layer sheets must be slightly wider than those of the first layer, because the tower is now just a little fatter.

Do not begin with a 31-inch-wide sheet, however, but with one just wide enough to span the space between the inner edge of the doubled door stud to the centerline of the adjacent single stud, or about 17¼ inches wide. You should have scrap pieces from the first course of siding which can be trimmed to fit. This strip must be completely painted on its underside (rough side) with glue, as must all the siding sheets of the second layer, and screwed tight at both vertical edges. Be sure to check for angular positioning before you finally attach it. The purpose of using this narrow strip is, of course, to arrange it so that subsequent seams will not occur over seams in the first layer. A second narrow strip will be required on the other side of the door. Once again, you'll have to cut around the landing joists, but it will be a little easier this time, as you don't have to cut so deeply into the sheets.

The second course of the second layer will be made up of

horizontal sheets exactly 40½ inches high, and about 91 inches long. Start at one side of the second-story door opening and, gluing and nailing as you go (screwing at sheet ends), work around again to the other side of the door opening. The top of this course should *not* be flush with the top of the second floor nailers, but exactly an inch below it, to allow for 1-inch-wide screen molding, to be applied later.

16. Now construct the third disk. Make it identical to the first disk, but as you did in making the second disk, leave the spacer blocks out and the upper layer off until you have it nailed onto the studs of the second story. For this operation you can safely stand on the second-story nailers. With the siding all on, the tower is now perfectly stable. When the disk is in place, glue and nail on the top layer of ¾-inch plywood, and glue and nail in the spacer blocks.

ROOF

It's time to build the roof, an eight-sided faceted one, which you will start with the kingpost. Before you make it, I suggest you go back and read the discussions of kingposts beginning on page 196 and have a look at the diagram and drawing of a kingpost for eight rafters on page 199.

Materials Needed

one 9-foot 2x6 for the kingpost
eight 5-foot 8-inch 2x6s for rafters
4 sheets of ⅝-inch plywood, exterior grade, for the underroof
4 pounds of 3½-inch common nails for framing
four 4½-inch common nails for framing, as called for
3 pounds of 2-inch serrated nails for the underroof
4 square feet of aluminum or galvanized flashing
a finial (optional)
2 bundles of roofing
2 pounds of roofing nails

Construction Steps

1. Cut a 9-foot 2x6 into three 3-foot segments. Don't worry about the little you will lose for kerf. Lay the three pieces tightly together, and then measure their sum thickness. If you have standard 2x6s, it should be about 4½ inches (see Dimensions of Boards, page 28), but get it as exactly as you can. Cut the segments to just that width, so that when the three pieces are laid together, they have a perfectly square cross section.

2. Glue the three timbers together, now, and clamp them well until the glue has set, but not yet fully hardened. Remove the clamps, then, and nail the pieces together. Use six 4½-inch nails, but three driven in from one side, three from the other. Then set the post aside until the glue has thoroughly dried.

3. Be very careful when you draw the cut lines on the post. Note that only a 6-inch area at the top of the post needs to be faceted, as that's where the rafters rest. Mark the octagon on one end and be sure all eight sides are identical in width. Also mark the lines down the sides which will delineate the new corners. Cutting the angled facets is most easily done with a power saw with a tilting blade, set at 45°, preferably a table or radial-arm saw. But a portable circular saw can also be used if you have a way safely to clamp down the post as you cut it. In a pinch you could even use a jigsaw, if you have a long enough blade for it. If you must do the job by hand, use a rip saw or a plane or, for that matter, a Surform, a course wood rasp, a chisel, or a drawknife. Don't go looking for a tool you don't have; a jackknife is enough if it's sharp, and if you've got a little time.

4. Center the kingpost upright in the exact center of the top disk, and draw a mark around it. It doesn't matter from the standpoint of strength where the rafters are, but you will probably want the area directly over the landing door to be free of rafters, so keep that in mind as you set the post. Remove the post, then, and edge the marked area with a rectangle of 2x4s, nailed in place. Be careful to nail into joists, not just into the plywood disk. Set the kingpost into the rectangle and toenail it into place with four 4½-inch nails, set around the post at 90° intervals. Drive the nails in at angles of about 45°, through the 2x4s of the rectangle, through the base of the kingpost, and into the disk.

5. It's rafter-cutting time. Start with eight lengths of 2x6, each 5 feet 8 inches long. At the end of one of them, make a 53° cut (a straight-across cut would be 90°). Hold that cut end against a flat at the top of the kingpost, while the other end of the rafter rests on the edge of the disk. When the top, sharp end of the rafter is even with the top of the kingpost, there should be a slight gap, ¼ inch or a little more, at the bottom *edge* of the rafter, at the kingpost. If there isn't, take the rafter down and trim the cut a little at a time until there is such a gap. Then make a mark ¼ inch beyond the edge of the disk, on the lower edge of the rafter. Lay the rafter down, now, and make a line starting at the mark, and 1½ inches long parallel to the 53° cut at the top of the rafter. Then, from the end of that 1½-inch line, make another line 2 inches long at 90° to it, running back to the edge of the rafter. In effect, you will have a 1½x2x2½-inch right triangle drawn on the rafter side, with the hypotenuse on the rafter edge. Lay the rafter up against the kingpost again, and see if it appears to you that if you cut that triangle of wood out, the rafter would rest on the cut that is true horizontal, with the true vertical cut right at the edge of the disk. Remember that the rafter will travel through a small arc as it drops from its uncut position to the cut one. If it doesn't look right, readjust the marking until it does—then cut it out. Test the rafter in place. If it isn't right, either trim it until it is, or if that isn't possible, start again with a new 5-foot 8-inch length of 2x6, putting into practice what you have learned with the first one.

When you have a good rafter, try it in different positions around the kingpost. If it fits in all positions, you know you can copy it exactly for the remaining seven rafters; if it doesn't, you either have the kingpost in the wrong place, or it isn't standing vertically, so adjust it until the rafter does fit all the way around. Produce seven other rafters, and trim the lower ends of all of them at 53°, parallel to the topcut—true vertical when in place.

6. Nail the rafters in place. Be sure each one is seated in the same way. Always follow one rafter with its diametrically opposed mate, so as to minimize strain on the kingpost. Two 3½-inch nails at the top and two at the saddle at the edge of the disk are sufficient for each rafter.

7. Before you apply any roofing, you should lay a strip of ¼-inch plywood around the edge of the upper disk, extending from the bottom edge of the disk up to the upper surface of the rafters. This will require a strip about 9¾ inches wide; you should be able to find the necessary pieces in your scrap. Cut notches in the upper edge for the rafters; you may also want to cut a shallow arc out of the upper edge between each pair of rafters, so the edge will meet the flat roof pieces neatly. Screw, nail, and glue the strip in place as you did the other siding. When it is in place, cut sixteen 4½-inch pieces of 1x2 (or 2x4, if you have some scraps you didn't use up as spacer blocks), and nail them vertically to the rafters, immediately behind the strip of upper siding, and then nail the siding to the blocks. That will help keep wasps from building their nests inside the roof, and provide a little extra rigidity.

8. For an underroof use triangular wedges of ⅝-inch exterior plywood. If you wish, curve the edge to conform to the curve of the building. Measure for each one carefully, to compensate for any misplacement of rafters. If you plan to have a finial at the top of the roof, trim the pointed ends of the wedges away, so that the underroof stops just at the kingpost, and when the wedges are all in place, measure for and fit an octagon of ⅝-inch ply that exactly fills the gap. If you don't want a finial, leave the points on, so that they meet at the center, above the kingpost. Don't worry about their short unsupported span at the top. Work caulking compound well into all the seams of the underroof.

9. The easiest finish roofing to apply is composition shingles. To cover an octagonally faceted roof, start at the bottom of each facet with a double layer, then proceed up the facet, trimming the ends at the ridge. Treat each facet like a separate roof face—which, of course, it is. At the top it is, obviously, impossible to nail on a double thickness, as you have only a tiny point. When the facets are all covered, run a row of single shingles up each ridge to the top, straddling it. If you mount a finial on the flat you have left at the top, then ring it with aluminum or galvanized sheet flashing that extends a short way up the side of the finial, and about 8 inches down over the shingles. If you don't use a finial, you can make a metal cap just like the predator-guard-cum-birdbath on page 44 of the Birdhouse chapter; just don't cut anything out in the middle. Always thoroughly caulk under all edges of flashing, and nail through the caulking. That means gobbing the stuff on pretty thickly, but that's the way to do it.

This would also be a good structure for using a thatched roof (see directions for thatching in the chapter on playhouses). If you use thatch, increase the pitch of the roof, though. A 45° pitch is about the minimum pitch that you can depend on if the thatch is to shed water reliably; you're better off with a 60° pitch. A

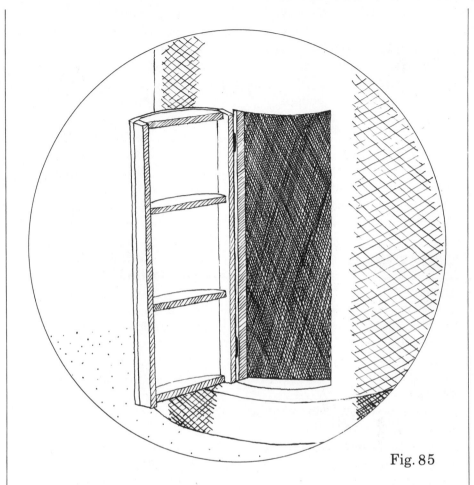

Fig. 85

pitch of 45° means, by the way, that the slope of the roof is at a 45° angle to the horizontal, that is, it rises one foot for every horizontal foot. A 60° pitch then makes the roof in cross-section an equilateral traingle, a triangle with three equal sides. If the diameter of the tower is 8 feet, then, not counting eave overhang, the roof will measure 8 feet from peak to wall.

DOORS

The two doors of the tower (upstairs and downstairs) should be made using horizontal shaped boards, like the stud braces.

The downstairs door, which in the plan does not have a window, should have four evenly spaced shaped horizontal pieces and two vertical 2x4 side pieces as shown in Fig. 85, page 219. The upstairs one will need the two side pieces plus three horizontal shaped pieces—one each at top and bottom and one at the bottom of the screening.

Sheathe the door in the same way you put on the siding, using the same ¼-inch plywood in two layers.

SCREENING

Materials Needed

To screen the second floor you will need:

25 running feet of 48-inch-wide copper or galvanized wire screening

280 feet of 1-inch-wide, ¼-inch-thick screen molding

¾-inch headed nails

¼-inch and ½-inch staples and a staple gun

Construction Steps

1. Run vertical strips of lath up the middles of all 10 doubled studs, covering the joint between the two 2x4s, from the top of the inner (1 inch higher) layer of siding up to the ring of siding around the uppermost disk. The distance spanned should be just about 46 inches, and be identical to the height of the window openings.

2. With a steel tape, measure around the circumference of the inner layer of siding, where it is exposed, immediately under the window opening, exactly from one vertical lath to another—inner edge to inner edge, not center to center. Cut a strip of molding just that length and tack it (with one ¾-inch nail, not quite driven home) to the stud by the door opening, butting against the side of the vertical molding strip, 16 inches up from the top of the nailer—which is now the window sill. Bend the strip around, keeping it horizontal, until you can tack the other end to the next stud, again butting against the vertical molding. The horizontal molding strip should bend with exactly—or nearly exactly—the curve of the tower. If it doesn't, trim it until it does, then nail it in place. Complement it, then, with another identical strip, running horizontally across the window opening, 16 inches below the top of the opening. Install similar bowed strips in the other eight window openings. Nail them in place at their ends with the ¾-inch headed nails. The final effect should be a double ring of molding strips extending all around the tower (except at

the door opening), and curved like the tower.

3. With the staple gun, tack one 48-inch-wide end of the roll of screening to the vertical molding strip next to the door opening. Space the staples no more than 3 inches apart, and be sure the end is kept vertical. Then, from inside the gazebo, unroll the screening, wrapping it around the outside of the window opening, until you reach the other side of the door opening. With the staple gun, temporarily tack the screening in place. Don't leave it slack around the tower, but don't stretch it very tightly just yet, either.

4. Starting at the door opening again, cut and nail molding over the first of the vertical strips. Space the nails about every 6 inches.

5. Now, working from a ladder outside the gazebo, stretch first the upper edge, then the lower edge, working from the fastened end, stapling as you go. The lower edge should be stapled to the exposed inch of under-layer of siding; the upper edge to the lower inch of the ring of siding above the window. As the opening is 46 inches high and the screening 48 inches wide, you should have an inch of overlap top and bottom. When you reach the first stud, stretch the screening tightly over the bowed strips, but not so tightly as to collapse them. Staple it in place, then lay a strip of molding over the vertical molding under the screening. In this fashion proceed around the tower until you reach the other edge of the door opening. This is one of those jobs, I'm afraid, that is easier described than done, but it's by no means impossible. Just take your time—and try to keep your temper.

6. Finally, cut molding strips to go over the bowed strips under the screen—sandwiching the screening as you have done with the vertical moldings. Tack them in place at the ends, and then, while someone holds a block of 2x4 or the like gently but firmly against the inside strip, to keep it from buckling in, staple the two bowed strips together, with screening in between, using ½-inch staples. Space the staples about every 3 or 4 inches.

FINISHING TOUCHES

Best to paint the tower now, before you put on a staircase that might get in your way. Before painting, caulk all the seams in the outer layer of plywood with exterior caulking compound. To best protect the wood, use an oil-base primer under an oil-base enamel. Paint the floors inside with oil-base deck paint. You will need 3 pounds of caulking compound and a caulking gun, 3 gallons of oil-base primer (for one coat), 4 gallons of oil-base enamel (for two coats), and 1 gallon of oil-base deck enamel for the floors (two coats).

STAIRS

You can climb upstairs on a simple ladder, or you can build an ordinary stairway—see page 144. Nicest, though, for this round tower is a round spiral staircase. Ready-made spiral staircases can be bought through your lumber dealer or builder's supply outlet, but it is also quite possible and not terribly difficult to make your own. For the design below you'll need either some experience in welding, and the necessary equipment for it, or you'll need the help of a professional welder.

The tools for the job (in addition to the welding equipment) may be rented from a rental company.

Tools needed
pipe vise on legs
two pipe wrenches suitable for
    3-inch inside diameters (i.d.)
pipe die for cutting pipe
    threads on 3-inch i.d. pipe
pipe-cutting tool for 3-inch pipe
    (better than a hacksaw
    because it makes a straight
    cut every time)
electric drill capable of drilling a
    1-inch hole

Materials Needed
11-foot length of black iron pipe,
    3 inches inside diameter
9-foot section of the same pipe
7-foot 6-inch section of the same
    pipe
3-inch flat pipe cap
eleven 3-inch i.d. T-unions
three 8-inch lengths of 1-inch
    (*outside* diameter) black iron
    pipe
eleven pyramid steel stair treads,
    cut as described below
cutting oil

Construction Steps
1. At the outer point of the stair landing to the second story, you made a small box frame with a hole 3½ inches square. Through this hole, drop a plumb bob (any heavy weight attached to a string will do) to find the spot on the ground directly below the hole. (By the way, beware of the effects of even a gentle breeze on the plumb bob.) Make the spot on the ground with a coin.

Measure the distance from the top of the landing to the ground. If you used the

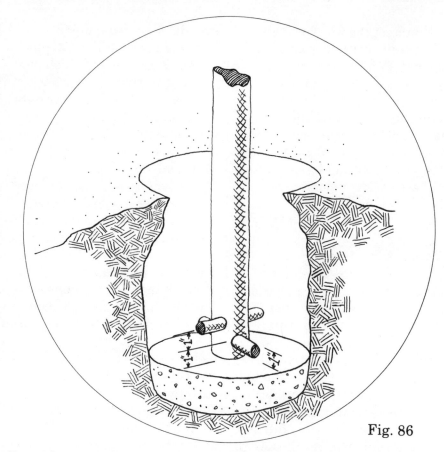

Fig. 86

dimensions given here and the ground is level, the distance will be 107½ inches; to simplify the discussion, we'll call it 108 inches—exactly 9 feet. (You, however, will have to deal precisely with whatever measurement you get.) The distance must be divided into an appropriate number of rises (see the section on stairs on page 144); twelve rises of 9 inches each would work well—eleven steel steps and the twelfth rise from the top step to the landing.

2. Directly below the hole in the landing, dig an 18-inch-deep hole in the ground with a post-hole digger. Pour in 3 inches of concrete, and let it set, Fig. 86.

3. Now cut the bottom section of the central post from the 11-foot piece of pipe. It is impossible to state the exact length you will need, as it all depends on the level of the concrete in the hole relative to the ground surface, the particular dimensions of the T-unions you get (they vary in size from different manufacturers) and the depth to which you are able to screw the pipe into the union. The important point is that the top of the bottom step should be exactly 8 feet 3 inches below the level of the second-story landing. Don't forget in measuring to take into account the thickness (¼-inch) of the stair tread and the pene-tration of the upper post end into the union.

4. When you have cut your pipe section to the correct length, drill a 1-inch hole through it, an inch from the bottom, and another, at right angles to the first, an inch higher up.

5. Push the 8-inch sections of small pipe through these holes to form a kind of double "T," and then place that end into the hole, resting on the concrete as in the illustration.

6. Holding the post vertical (use a level), pour the hole full with concrete, and let it set. It would be helpful if you were to get hold of a smaller pipe, just small enough to fit inside the pipe in the concrete, and tall enough to reach up through the boxed hole in the upstairs

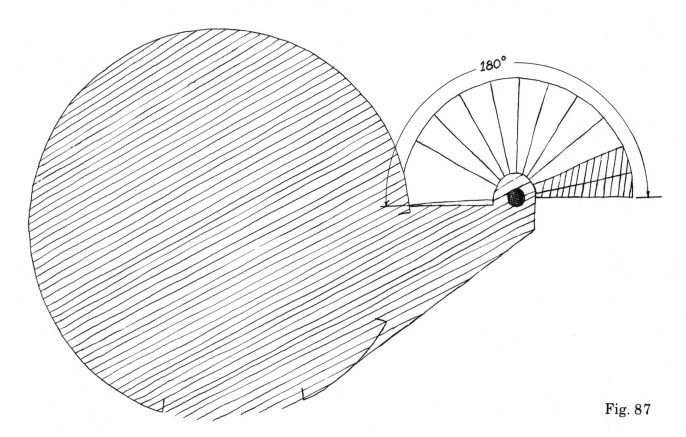

180°

Fig. 87

landing. That way you could be sure you were off to the right start with your stairs.

7. Cut the 9-foot and the 7½-foot pieces of pipe into eleven 18-inch segments. Cut threads on one end of each segment and screw each one into the side hole of a T-union, that is, the hole representing the vertical stroke of the T. These are the bars that will support the stair treads.

8. When the concrete has set, cut threads on the protruding end of the pipe set in the ground. Don't cut more than an inch and a quarter, and screw onto it one of the T-unions, so that the 18-inch pipe segment sticks out at right angles to the vertical pipe. Don't worry about the angle at which the joint tightens for the time being.

9. From this point, the whole staircase just screws together. There should be exactly 9 inches of rise from the top of one stair support pipe to the top of the next, so cut your sections of center post accordingly, never forgetting to allow for the penetration of the post into the unions, top and bottom, see Fig. 88, page 224. Be careful also not to cut the threads too far down each pipe segment. A union should begin to tighten when the top of its support pipe is between 9¾ and 9½ inches above the top of the support pipe below. In all there are eleven iron steps and the top landing, for a total of twelve rises.

10. At the top of the center post is a pipe section, about 3½ feet long, screwed into the topmost T-union. That section protrudes through the box at the end of the stair landing, and is capped at the top. It functions as a hand-hold when going up or down the stairs—an upstairs newel post—and, after you do a little work to it, it will support the outer end of the landing. For the moment, however, leave it screwed hand tight, and unattached to the landing.

11. It's time now to decide on the amount of rotation you want the whole spiral staircase to have see Fig. 87, page 223. In order to avoid banging your head on a step above, you cannot have a rotation from bottom step center to top landing edge of more than 270°—that is, three-quarters of a full turn. If you want this maximum spiral, you will be able to have nice, wide steps—23.5° per step, which means almost 10 inches wide at the end, and about 6½ inches wide two-thirds of the way to the end, about where people would normally step. That's almost as deep as some home stair treads. The disadvantage is that the

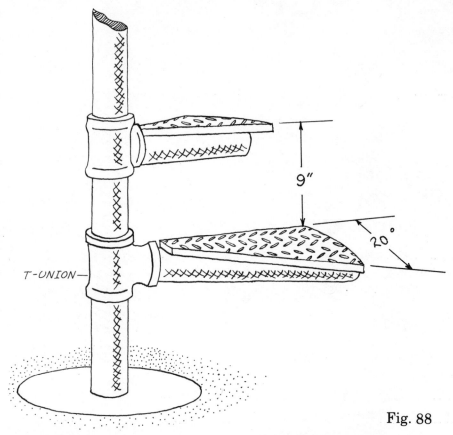

T-UNION—

Fig. 88

bottom step will face directly at the wall of the tower, less than 2 feet away. A much nicer placement for the bottom stair is with its center 180° rotated from the top landing edge. That way the bottom step will face in the same direction as the downstairs door, which is nice, but the rotation for each step is limited to about 16°, or about 6½ inches at the end, and about 4⅓ inches where people would step. That's

not as bad as it sounds, though, as you can—and should—make the treads themselves larger than the angle of rotation, if you decide on 180° total spiral. An arc of 20° would be a good compromise, yielding a step width of 8½ inches at the end and 5⅔ inches where people would step, see Fig. 88.

12. To determine whether two adjacent pipes are 16° apart (or 23½°, or anything in between you might decide on), you need a gauge, which you can easily make from a scrap of plywood 2 feet long and a foot wide. Using a pencil, string, and a nail, draw an arc at one end of the scrap with a radius of 2 feet. Then, using the nail-hole as the center point, with an ordinary school protractor mark out the angle—not of the stair tread you plan on, but of the rotation from step center to step center, or 16° if you plan on the 180° option. Cut the pie-piece-shaped wedge out, and then cut exactly enough from the point to place the curve 2 feet from the center of the central pipe when you rest the gauge on a support pipe, its inner end butted against the flange of the T-union. In effect you have made a dummy step, although somewhat narrower than the steps you actually plan. With this gauge and a carpenter's square you will be able to measure the rotation between step support pipes. Use a level to be sure you have the gauge and square straight each time.

13. Now you can begin screwing things tight. Begin with the joint between the bottom T-union and the pipe base set in the concrete. Tighten that joint until the bottom support pipe is exactly at the angle you want. If you have opted for the 180° spiral, then the pipe should be pointing exactly away from the center of the tower.

When you have the support pipe set correctly, tack-weld it in place, but make the weld minimal, so you can break it if you find you've made a miscalculation. Use the gauge, square, and level, and set and tack-weld the remaining T-union joints.

When the support pipes are all in and at the correct angles, reinforce the welds at each joint, and with a large pipe wrench tighten in the support pipes themselves, and weld the joints there, too. You wouldn't want those joints to begin to unscrew sometime and rotate in a possibly perilous way.

14. Now that everything is tightened down, climb up the support pipes and drill a 1-inch hole through the center post immediately below the landing. If you can, wedge the landing up a little with a 2x4 from the ground before you make the hole. Put the last 8-inch segment of 1-inch (O.D.) pipe through the hole, and let the landing rest on it. Weld the pipe in place. If necessary, you can interpose shims between the pipe and the landing until the landing is properly supported.

15. The last step, of course, is to make up and weld in place the stair treads. Unless you are a skilled metalworker, you will probably find it best to have the treads made up by a blacksmith or ironworker from ¼-inch patterned sheet plate. Have them made exactly as you made the gauge, but instead of having an angle of 16°, make them 20°, assuming you elected the 180° option. As with the gauge, the inner edge of the tread should abut against the flange of the T-union, and the outer curve should be exactly 2 feet from the center of the central pipe.

The plan for the staircase, as presented, does not include an outside railing. Instead the central post is continued up to the level of the railing on the second floor landing, and above that by 3 feet, and is intended to be used as a hand-hold. Still, building codes in some areas may require that there be an outside railing, in which case you will have to get a blacksmith or ironworker to make one for you.

When the staircase is finished, go over the whole thing with a metal file and take off the rough edges. There will be a lot of them. The more rounded all the edges are, the less likely someone will get a nasty poke or scratch.

Glossary

auger A tool for drilling holes in wood, consisting of a *bit* with a spirally-inclined plane and a handle. See *brace and bit*.

baluster A rod supporting a stair or porch railing.

batten A strip of wood, or a board, used to bridge or seal a gap between two other boards, or to hold two or more boards together.

bit That part of a drill or *auger* which actually makes the hole, as separate from the handle, and the *chuck*, which holds the *bit* in the handle.

board A piece of dimension lumber nominally 1 inch thick and 3 or more inches wide.

box nail A thin-shanked, flat-headed nail (see photo, page 64).

brace and bit A tool for making holes in wood, consisting of a cranked handle with a *chuck*, and a *bit*.

bridge A brace between two *joists* or two *rafters* to keep either from twisting or splaying.

bubble level See *level*.

bundle The standard roofing unit. One bundle covers 33 square feet.

butt The end of a piece of wood.

butt nail See *toenail*.

carriage bolt A round-headed bolt for wood. Underneath the round, domed head is a short square section of shank. As the nut is drawn tight, the square section is drawn into the wood and prevents the head from turning.

casement window A window hinged at the side, swinging either in or out.

centers As in the phrase, "on 16-inch centers." The measured distance between the longitudinal center of one stud (or joist, rafter, etc.) and its neighbor. Differs from "set 16 inches apart" in that the latter refers to the distance between adjacent sides.

chalk line A string coated with chalk dust, used by carpenters to mark long, straight lines. The string can be rubbed on blue **carpenter's chalk** or unwound from a case with chalk dust, and is then stretched taut over the surface to be marked, drawn slightly away from the surface, and then allowed to snap back, leaving a line of chalk deposited along its length.

chip board See *particle board*.

chisel A steel tool with a wedgelike blade at one end which is used with a hammer to shape or gouge wood.

chuck A cylindrical clamp for holding a *bit* securely in a drill handle.

cleat A small strip of wood nailed horizontally against the inside of a stair stringer to support a tread. Any piece or strip of wood fixed directly to another piece of wood to brace it or support another board.

common nail A flat-headed nail with a heavier head and thicker shank than a *box nail* (see photo, page 64).

compass Any device for drawing a circle.

compass saw A small handsaw

with a narrow, tapering blade, for cutting curves. Similar to but smaller than a *keyhole saw*.

compound angle cut A cut that is at other than a right angle to the length of a board or timber when viewed *both* from the wide and the narrow sides of the board or timber.

countersink To drive (or screw) a nail (or screw) until the head is below the surface of the wood. The cavity above the head may then be filled with putty or wood filler and the nail or screw thus concealed. Also, a tool for drilling a cavity for a screw head, or a tool for setting a nailhead below surface level.

d (as in **6d** nail) is the old English symbol for **penny**, and stands for Latin *denarius*. A couple of centuries ago nails were sold at so many pennies a gross (144). The bigger the nail, the more you had to pay. We still grade nails by size this way. The **d** is pronounced "penny."

dimension lumber Wooden boards or timbers sawn to specific dimensions, such as two-by-fours, two-by-sixes, etc. (see *dressed* and *full dimension*).

double-hung window A window in two sections, both vertically movable, one above and slightly overlapping the one below. The usual American house window. Also called a **sash window.**

double strength One of two standard thicknesses of window glass, about $3/32$ or $7/64$ inch thick. Called **double diamond** in Canada. (See *single strength*).

dressed A term applied to dimension lumber that has been planed smooth. Dressed lumber, which is standard building lumber, measures from ¼ to ⅝ inch less in width and thickness than its *nominal dimensions*, that is, a dressed 2x4 will actually measure only about 1½ x 3½ inches. There appears to be a tendency now for lumbermills to increase the disparity between nominal and actual dressed dimensions. (See *full dimension*.)

eave That part of a roof that extends beyond the top of the outside wall, usually along the lower edge of a roof at the side of a building, but may also refer to the part of the roof that extends beyond the wall on a gable end.

eave height The height of a side, rafter-supporting wall.

fascia board A board nailed to the lower ends of rafters to give the edge of the roof a finished look. Also used sometimes along the end rafter of gable ends.

finial A vertical ornamental termination, usually round in section. May extend upward, as from a conical roof, or hang downward, as from a piece of under-eave scroll work.

finish [flooring, roofing, siding, etc.] The final layer.

finishing nail A headless or nearly headless nail that holds entirely by friction. It may be countersunk and concealed with putty or wood filler.

flashing Sheet metal used to cover a roof peak or the joint of a roof with a wall or chimney, etc. Usually the last element of a roof to be applied.

flathead screws Screws that have heads flat on top— the standard shape for wood screws.

flush Even with, at the same level with.

framing square A carpenter's square, usually with sides 18 and 24 inches long, specially marked for cutting rafters.

full dimension Refers to dimension lumber which

actually has its *nominal dimensions*. A full-dimension 2x4 actually measures 2 inches by 4 inches. Full-dimension lumber is generally *rough cut*.

furring strip A strip of wood secured to a surface into which nailing is difficult, to which other building lumber is in turn nailed.

gable The end point of a peaked roof, usually including the triangular section of wall below the peak.

gable-end An end of a building on which there is a gable.

gambrel roof A gable roof, each side of which has a steep slope below a shallower slope.

girdle board See *skirting board*.

gusset A flat brace joining two abutting boards or timbers, often a triangle of plywood joining two rafters at the roof peak.

hardboard A generic term; the trade name *Masonite* is commoner. A brown board ⅛ to ⅜ inch thick made of pressed sawdust and glue, usually sold in 4x8-foot sheets, smooth on one side, rough on the other. *Tempered hardboard* is darker, stronger, and smooth on both sides.

header A horizontal (frequently doubled) timber over a window or door opening.

hole cutter When I suggest using a hole cutter, I have in mind a cylindrical saw which comes fitted in a nest of about eight siblings of varying sizes, all clustered around a ¼-inch drill bit. These sets of hole cutting saws can be bought quite cheaply where tools are sold.

house-frame construction Describes the standard, generally accepted ways of putting timbers and boards together to make a wood-frame house, common elements of which are 2x8-inch floor joists, 2x4-inch wall studs and 2x6-inch rafters, all set 16 inches (sometimes 24 inches) apart. There are many variations.

jamb The inner framework of a door or casement window frame facing the edge of the door or window. Usually there is a jamb on both sides and the top of the frame, and a *sill* on the bottom. The narrow molding on the jamb that stops the door is called a *stop* or *stop strip*.

jigsaw A saw with a narrow, short blade mounted in a deeply recessed frame, suitable for cutting curves. May also be

called a *coping* or *scroll saw*, or especially in its powered form, a *saber saw*.

joist A horizontal timber supporting a floor.

joist carrier Can be either a wooden *cleat* supporting the *joist* at the *perimeter board*, or a metal sling that hangs from the perimeter board and cradles the joist.

kerf Whenever you saw wood, a certain amount of length is lost to the thickness of the cut itself. That loss is called kerf.

keyhole saw A handsaw with a thin, tapering blade, often used to start a cut from a hole drilled in a board or building sheet with a *brace and bit* or an *auger*.

lag screw A large and heavy wood screw with a square or hexagonal head used to join large boards or timbers. Also called a **lag bolt**.

lath A thin strip of wood, usually rough-cut, typically ¼ inch thick by 1¼ inches wide. May also refer to a number of laths used as an underpinning for plaster.

lattice A thin strip of wood, usually ¼ inch thick, distinguished from *lath* by being planed smooth, and available in widths from 1 to 3½ inches.

level A device used by carpenters to find true horizontal and vertical. This usually is a straight piece of wood or metal into which are set slightly curved glass tubes containing colored alcohol with air bubbles in them. When the bubble is centered between two marks on the tube, the level is horizontal or vertical. Also called **spirit** or **bubble level.**

machine bolt A bolt threaded to take a nut and which has a square or (more commonly) hexagonal head. Differs from a **machine screw** in that the latter has a round head slotted for a screwdriver.

Masonite See *hardboard.*

miter cut A precise cut, usually at an angle other than 90°, made at the end of a strip of wood, to be joined to another similarly cut strip of wood, as the pieces of a picture frame. Usually made in a **miter box** with a **miter saw** (also called a **backsaw**), but some power saws (most commonly radial-arm saws) can be adjusted to make miter cuts.

molding A long, decoratively shaped strip of wood.

mortise A slot or groove in a piece of wood cut to receive a *tenon* or *tongue* cut into another piece of wood, to which the first piece is to be fastened. Also the slot in the edge of a door cut to hold a lock or latch mechanism.

nailer A horizontal timber, usually a 2x4 set between two studs, both as a brace for the studs, and to provide something between the studs to which siding can be nailed.

newel The main post at the upper or lower end of a stair banister. Also the center post of a spiral staircase.

nominal dimension The named dimensions of lumber, such as 2x4, usually larger than the actual *dressed* measurements.

pan-head screw A screw with a flattened round slotted head, most commonly a *sheet-metal screw.*

particle board A wood product made of wood chips, sawdust, and glue, available in many of the same thicknesses as plywood, in 4x8-foot sheets, and usable, like plywood, for underfloors, underroofs, and sheathing. Also called **chip board.**

perimeter board The heavy board (usually 2x10 or 2x12) that forms the base perimeter of a building, above the foundation, and to which the first-floor *joists* are nailed. In pole buildings called a **skirting** or **girdle board**.

pitch The angle of a roof, usually expressed in feet of vertical rise for feet of horizontal run. A "one for one" roof would thus have an angle of 45°. Also used with staircases.

plane A hand tool for smoothing and occasionally for shaping wood, consisting of a wooden or metal block with a flat, smooth sole, through which slightly protrudes at an angle a flat, straight blade.

planed Smoothed with a plane. Used of lumber, it means the same as *dressed.*

plate A horizontal, frequently doubled timber forming the top of a wall. Sometimes also used to carry the same meaning as *shoe,* but then distinguished as **base** or **sole plate** at the bottom and **top** or **head plate** at the top.

plumb bob A weight attached to a string for finding a vertical line.

pole-frame construction A

method of building wooden buildings which starts with poles set vertically into the ground, usually at 4- or 8-foot intervals. More suited to sheds and outbuildings than dwellings.

posthole digger A tool (as used in this book; not a person) for digging postholes. As a hand tool, it consists of two semi-cylindrical spades, each with its own long handle, and hinged together at the tops of their blades, which form then a cylinder capable of opening and shutting its jaws, below. As a power tool, it is a large auger bit driven by a small gasoline engine.

proud As used by carpenters, means slightly raised or elevated above a surrounding level. Not *flush* nor *countersunk*.

purlin A longitudinal roof timber which supports rafters below the roof peak, but above the eaves. Frequently used at the gambrel of a *gambrel roof*.

quarter-round A common molding, with two flat sides and a 90° curved one. Often used where floor and baseboard come together.

rafter A timber extending from roof peak to wall top and slightly beyond, to which the underroof is nailed. Rafters are to the roof what joists are to the floor—the basic supporting timbers.

ridge pole A timber (usually squared, not round as the word *pole* might suggest) that extends along the peak of a roof, and against which the upper ends of the rafters are butted. An optional element of house-frame or pole-frame construction.

rise See *pitch*.

riser The vertical board separating two adjacent *treads* of a staircase.

roofing nail A galvanized (zinc-coated), short nail with a broad, flat head for nailing shingles, tarpaper, etc., to a roof (see photo, page 64).

roof pitch See *pitch*.

rough cut Not planed or smoothed; as sawn by the lumber mill's log saw. May also mean *full dimension*.

rough opening The space left for a window or door or the like in the skeletal frame of a building.

run See *pitch*. May also refer to the lengh of a wall.

saber saw See *jigsaw*.

sash window See *double-hung window*.

scaffolding nail A double-headed nail , as if a common nail had a secondary head about ¼ inch below the real head. It can be pounded in firmly, as to hold a scaffolding, but is easy to remove.

screed A tool for **screeding**, that is, for working air bubbles out of concrete and settling it before it hardens. May also refer to a tool for leveling wet concrete.

screened rock At the quarry, crushed rock is shaken through progressively finer wire mesh screens. When a rock can no longer pass through a further screen, it is graded at that level. If it stops at the 1-inch mesh, for example, having passed through the 1¼-inch screen, it is graded as 1-inch screened rock.

scroll saw See *jigsaw*.

septic system A system for disposing of toilet waste, including a holding tank (septic tank) and perforated pipes (laterals) leading from it, through which the liquid sewage soaks into the ground. The sludge remaining in the tank must be pumped out from time to time.

shake A thick wooden shingle; thin ones are called **shingles**.

sheathing An exterior under-covering, usually for walls or roofs, commonly plywood or chip board, over which is nailed the finish roofing or *siding*.

sheet-metal screw A screw, usually with a pan head and made of very hard steel or alloy, able to cut its own thread in a hole drilled in sheet metal.

shim A thin piece of wood used to wedge two framing elements tightly together or apart. Window and door units are usually **shimmed** tightly into the *rough openings*.

shiplap A board with a nominal thickness of 1 inch stepped along both edges, so that adjacent shiplap boards may fit together and overlap slightly (usually by ¼ inch). See Fig. 55c on page 151; compare *tongue-and-groove*.

shoe A horizontal timber forming the bottom of a wall. Also called *sole plate*.

side rail See *stringer*.

siding Any covering fixed to the outer frame of a building to keep out the weather. Unlike the interior finish wall, the siding is always applied to the outside of the frame wall, but may be put on over a rough underlayer called *sheathing*.

siding nail A thin nail with a thin, flat head, used for fastening siding to the wall frame. Often serrated or threaded on the shank so as to hold better in the wood.

sill A board or piece of molding forming the bottom of a door or window frame; see *jamb*. Also a board bolted flat on top of a foundation as a base for *perimeter boards*, etc.

single strength One of two standard thicknesses of window glass, about ⅛ inch thick. Called **single diamond** in Canada. (See *double strength*).

skid A longitudinal foundation board for a small building, like a little shed or a doghouse. The skid will take the place of a stone or concrete foundation and is not sunk into the ground. If necessary, the building may be pulled along on its skids.

skirting board A perimeter of wide boards, commonly 2x10 or 2x12, fixed to the inside of a pole building's poles and used to contain a gravel or poured concrete floor, or to support floor joists. Sometimes called a *girdle board*. May also carry the same meaning as *perimeter board*.

sole plate See *shoe*.

span The unsupported distance bridged by a board, usually a joist or rafter, or the distance between two studs.

spike A long (6 inches or more) relatively thin, sharp nail with a broad head, like an oversized *box nail*. For nailing logs or posts together.

spirit level See *level*.

square (1) A carpenter's tool for marking building materials at right angles to their length, or for checking that two construction members are actually at right angles to one another. Carpenter's squares usually have the shape of a capital L.

square (2) Roof areas are measured in squares, which are 100 square feet each. It takes three bundles of roofing to cover one square.

stringer A timber (usually 2x4) linking the far ends of a pair of rafters, forming a triangle. Also the side member of a staircase (when it is also called a **side rail**). The *treads* and *risers* are fixed between the two stringers.

stud A vertical timber in a wall frame.

stop molding The molding or wood strip against which the inner edges of a door or casement window press when shut. Also called **stop strip** or simply **stop**.

stove bolt Similar to a *machine screw*, but with a coarser thread.

subfloor The flooring laid on the joists, under the finish floor.

Surform A patent combination rasp and plane with replaceable blade, made by the Stanley Company of New Britain, Connecticut. A very handy wood-shaping tool.

tenon A narrow, protruding rectangle of wood, cut as a part of a larger piece. The tenon fits into a *mortise* cut into another board. The mortise and tenon are usually then glued together, and thus join two larger construction members.

timber Any piece of dimension lumber not a board, including all nominally 2-inch-wide lumber, posts, etc.

toenail To drive nails diagonally through the sides of a butt-end of a timber as a means of fixing the timber to another at its butt, when it is impossible to nail through the second into the butt of the first.

tongue Like a *tenon*, but extending the whole length of the side of a board or timber; a rectangular ridge on one board or timber designed to fit tightly into a groove in the next board or timber. A similar joint to *shiplap*, but tighter.

tongue and groove See *tongue*.

tread One of the flat, horizontal boards on which one steps, on a staircase.

Vise Grip wrench An adjustable, locking plier made by the Peterson Company of DeWitt, Nebraska, in various sizes. One of the handiest tools ever invented.

Index